Functional Polymers in Sensors and Actuators

Functional Polymers in Sensors and Actuators

Fabrication and Analysis

Editors

Akif Kaynak
Ali Zolfagharian

MDPI • Basel • Beijing • Wuhan • Barcelona • Belgrade • Manchester • Tokyo • Cluj • Tianjin

Editors

Akif Kaynak
School of Engineering,
Deakin University, Geelong
Australia

Ali Zolfagharian
School of Engineering,
Deakin University
Australia

Editorial Office
MDPI
St. Alban-Anlage 66
4052 Basel, Switzerland

This is a reprint of articles from the Special Issue published online in the open access journal *Polymers* (ISSN 2073-4360) (available at: https://www.mdpi.com/journal/polymers/special_issues/ Polymers_Sensors_Actuators).

For citation purposes, cite each article independently as indicated on the article page online and as indicated below:

LastName, A.A.; LastName, B.B.; LastName, C.C. Article Title. *Journal Name* **Year**, *Article Number*, Page Range.

ISBN 978-3-03936-868-6 (Hbk)
ISBN 978-3-03936-869-3 (PDF)

Cover image courtesy of Ali Zolfagharian.

Contents

About the Editors

Akif Kaynak received his BSc degree from the University of Manchester in the UK, his MSc degree from Rutgers State University of New Jersey, USA, and PhD from the University of Technology, Sydney (UTS), in Australia. He is a leading researcher in stimuli-responsive polymers with soft actuators application within the School of Engineering, Deakin University, Australia. He has more than 130 publications, a book chapter, a book on conducting polymers and citations exceeding 3000, and an H-index of 29. He is a regular reviewer for various international journals and part of the advisory board of Sensors journal. He guest-edited issues on stimuli responsive polymers in Materials and finite element methods in smart materials and systems in Polymers, MDPI.

Ali Zolfagharian is an Alfred Deakin Medalist for Best Doctoral Thesis and Alfred Deakin Postdoctoral Fellowship Awardee, at Deakin University, Australia. He is a Mechanical Engineering lecturer in the School of Engineering, Deakin University, Australia. Dr. Zolfagharian is one of the foremost researchers in Australia in the 3D/4D printing of soft robots and soft actuators. He has thus far received $206k funding from 3DEC (Deakin Digital Design and Engineering Centre), IISRI (Institute for Intelligent Systems Research and Innovation), and an industrial firm. His recent research outputs in the field of 3D and 4D printing include the guest editing of four special issues in Polymers, Materials, Applied Sciences, one edited book, and 46 articles.

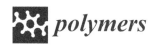

Editorial

Functional Polymers in Sensors and Actuators: Fabrication and Analysis

Akif Kaynak * and Ali Zolfagharian *

School of Engineering, Deakin University, Geelong, Victoria 3216, Australia
* Correspondence: Akif.kaynak@deakin.edu.au (A.K.); a.zolfagharian@deakin.edu.au (A.Z.)

Received: 3 July 2020; Accepted: 7 July 2020; Published: 15 July 2020

Keywords: functional polymers; sensors; actuators; 3D printing; 4D printing

Recent advances in fabrication techniques have enabled the production of different types of polymer sensors and actuators that can be utilized in a wide range of applications, such as soft robotics, biomedical, smart textiles and energy harvesting. This Special Issue focuses on the recent advancements in the modeling and analysis of functional polymer systems.

The first paper published in the issue presents the work of Yu and colleagues in Shandong University, China, in which they synthesized hydrogel materials that could respond to the surrounding environment by a color change inspired by nature [1]. The researchers have presented an efficient method to improve the photonic sensing properties of polymeric gels by using non-close-packed monolayer colloidal crystals as the template. The authors developed an ultrathin photonic polymer gel film which exhibited significant improvement in responsiveness and linearity towards pH sensing compared to those prepared earlier, achieving fast visualized monitoring of pH changes with excellent cyclic stability and a small hysteresis loop.

In the second article, a collaborative research team, including the University of Tartu, Estonia and Ton Duc Thang University, Vietnam, developed software for driving and measuring ionic electroactive material-based systems [2]. A set of functions, hardware drivers, and measurement automation algorithms were developed in the National Instruments LabVIEW 2015 system to control synchronized isotonic (displacement) and isometric (force) measurements over a single compact graphical user interface called electro-chemo-measurement software (IIECMS). The suitability of the proposed software was successfully tested on the two different materials representing high stress, strain and low strain characteristics.

The Special Issue progresses to the third manuscript with the work of Park and colleagues from Inje University, South Korea, on ionic electroactive polymer actuators (IEPAs) which are interesting for their flexibility, lightweight composition, large displacement, and low-voltage activation [3]. They have developed a graphene oxide–silver nanowire (GO–Ag NW) based IEPA with Triton X-100 nonionic surfactant to transform the PEDOT:PSS capsule into a nanofibril structure. The fabricated actuator in this work showed improvements in stability, electrical conductivity, and driving performance.

In the fourth article, in an international collaboration, Wang from Donghua University and other colleagues from Texas Tech University and California State University, Fullerton, developed a 3D-printed wearable strain sensor with promising conductivity and transparency suitable for healthcare and soft robotics applications [4]. They combined agar and ionic thermo-responsive alginate to improve the shape fidelity of the hydrogel for 3D printing. With the addition of agar, the rheological characteristic of the 3D printing ink was enhanced for precision printing. In addition, alginate was used to improve the mechanical characteristics of the sensor to a level required for the so-called "electronic skin".

The researchers in Prince of Songkla University presented the fifth article investigating electrostrictive polymers with applications in biomedical sensors, actuators and energy harvesting devices [5]. The authors worked on increasing the dielectric properties and microstructural β-phase in the poly(vinylidene fluoride-hexafluoropropylene) (P(VDF-HFP)) by optimizing electrospinning conditions and thermal compression. The high electrostatic field in the electrospinning process caused orientation polarization, which helped transform the non-polar α-phase to electroactive β-phase in the formed fibers. Additionally, the increase in compression temperature of up to 80 °C resulted in an increase in the crystallinity and the dielectric constant. The results showed the efficacy of the proposed method to improve electrostriction behavior based on the dielectric permittivity and interfacial surface charge distributions for applications in actuator devices, textile sensors, and nanogenerators.

The seventh contribution to the Special Issue focused on the bending problem of a piezoelectric cantilever beam via theoretical and experimental methods. Due to the extensive applications of piezoelectric polymers in the design of intelligent structures, including the sensors and actuators, Yang and associates from Chongqing University, proposed a method to deal with the challenge of solving the governing equations of these materials due to the force–electric coupling characteristics [6]. To do so, they derived the theoretical solution of the bending problem of piezoelectric cantilever beams by the multi-parameter perturbation method, which is a general analysis method for solving approximate solutions of non-linear mechanical problems. The solution of their proposed method was successfully validated with the experimental results as well as existing solutions in the literature.

International researchers from Nottingham Trent University, University of Tehran, Deakin University, and University of Glasgow conducted a new level of study in digital fabrication, publishing their findings in harnessing variable bandgap regions by 4D printing via shape-adaptive metastructures [7]. Focusing on how four-dimensional (4D) metastructures could filter acoustics and transform filtering ranges, the authors used fused deposition modeling (FDM) printing with a single printer nozzle to experiment with shape memory polymer (SMP) materials with self-bending features. Additionally, the mechanism for the creation of metastructures capable of manipulating elastic wave propagation to find bandgaps was demonstrated. The authors claim that the state of the art 4D printing unlocks potentials in the design of functional metastructures for a broad range of applications in acoustic and structural engineering, including sound wave filters and waveguides.

The eighth contribution to the issue is a review of polymer-based microelectromechanical systems (MEMS) electromagnetic (EM) actuators and their implementation in the biomedical engineering field written by a national collaboration of Yunas and colleagues among three universities in Malaysia [8]. The study highlighted the recent development of electromagnetically driven microactuators in terms of the materials, mechanism of the mechanical actuation, and the state of the art of the membrane developments for biomedical applications, such as lab-on-chip and drug delivery systems. The authors envisaged that the polymer composites will eliminate the need for a conventional bulky permanent magnet in electromagnetic actuators in the near future.

This issue finalizes with the work of researchers from the School of Engineering at Deakin University on the development of the wearable strain sensors [9]. In this work, an electrically conductive dynamic hydrogel was designed and produced by incorporating ferric ions and tannic acid-coated chitin nanofibers (TA-ChNFs) into the hydrogel network. TA-ChNFs had a reinforcing role as nanofillers and also acted as dynamic cross-linkers, thus imparting an outstanding self-healing ability and high strength to the hydrogel. Moreover, the hydrogel displayed excellent stability with repeatable self-adhesive properties, with the ability to attach to almost any surface. This electroconductive and tough hydrogel with autonomous self-healing and self-recovery properties appeared to be an excellent candidate for wearable strain sensing devices.

Funding: This research received no external funding.

Acknowledgments: As the Guest Editors we would like to thank all the authors who submitted papers to this Special Issue. All the papers submitted were peer-reviewed by experts in the field whose comments helped improve the quality of the edition. We would also like to thank the Editorial Board of Polymers for their assistance in managing this Special Issue.

Conflicts of Interest: The authors declare no conflict of interest.

Polymers **2020**, *12*, 1569

References

1. Yu, S.; Dong, S.; Jiao, X.; Li, C.; Chen, D. Ultrathin Photonic Polymer Gel Films Templated by Non-Close-Packed Monolayer Colloidal Crystals to Enhance Colorimetric Sensing. *Polymers* **2019**, *11*, 534. [CrossRef]
2. Harjo, M.; Tamm, T.; Anbarjafari, G.; Kiefer, R. Hardware and Software Development for Isotonic Strain and Isometric Stress Measurements of Linear Ionic Actuators. *Polymers* **2019**, *11*, 1054. [CrossRef]
3. Park, M.; Yoo, S.; Bae, Y.; Kim, S.; Jeon, M. Enhanced Stability and Driving Performance of GO–Ag-NW-based Ionic Electroactive Polymer Actuators with Triton X-100-PEDOT:PSS Nanofibrils. *Polymers* **2019**, *11*, 906. [CrossRef] [PubMed]
4. Wang, J.; Liu, Y.; Su, S.; Wei, J.; Rahman, S.E.; Ning, F.; Christopher, G.; Cong, W.; Qiu, J. Ultrasensitive Wearable Strain Sensors of 3D Printing Tough and Conductive Hydrogels. *Polymers* **2019**, *11*, 1873. [CrossRef]
5. Tohluebaji, N.; Putson, C.; Muensit, N. High Electromechanical Deformation Based on Structural Beta-Phase Content and Electrostrictive Properties of Electrospun Poly(vinylidene fluoride- hexafluoropropylene) Nanofibers. *Polymers* **2019**, *11*, 1817. [CrossRef]
6. Yang, Z.-X.; He, X.-T.; Jing, H.-X.; Sun, J.-Y. A Multi-Parameter Perturbation Solution and Experimental Verification for Bending Problem of Piezoelectric Cantilever Beams. *Polymers* **2019**, *11*, 1934. [CrossRef] [PubMed]
7. Noroozi, R.; Bodaghi, M.; Jafari, H.; Zolfagharian, A.; Fotouhi, M. Shape-Adaptive Metastructures with Variable Bandgap Regions by 4D Printing. *Polymers* **2020**, *12*, 519. [CrossRef] [PubMed]
8. Yunas, J.; Mulyanti, B.; Hamidah, I.; Mohd Said, M.; Pawinanto, R.E.; Wan Ali, W.A.F.; Subandi, A.; Hamzah, A.A.; Latif, R.; Yeop Majlis, B. Polymer-Based MEMS Electromagnetic Actuator for Biomedical Application: A Review. *Polymers* **2020**, *12*, 1184. [CrossRef]
9. Heidarian, P.; Kouzani, A.Z.; Kaynak, A.; Zolfagharian, A.; Yousefi, H. Dynamic Mussel-Inspired Chitin Nanocomposite Hydrogels for Wearable Strain Sensors. *Polymers* **2020**, *12*, 1416. [CrossRef] [PubMed]

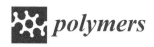

Article

Ultrathin Photonic Polymer Gel Films Templated by Non-Close-Packed Monolayer Colloidal Crystals to Enhance Colorimetric Sensing

Shimo Yu, Shun Dong, Xiuling Jiao, Cheng Li * and Dairong Chen *

National Engineering Research Center for Colloidal Materials and School of Chemistry and Chemical Engineering, Shandong University, Ji'nan 250100, China; 201411530@mail.sdu.edu.cn (S.Y.); shundong1995@mail.sdu.edu.cn (S.D.); jiaoxl@sdu.edu.cn (X.J.)
* Correspondence: chengli@sdu.edu.cn (C.L.); cdr@sdu.edu.cn (D.C.); Tel.: +86-531-88364280 (C.L. & D.C.)

Received: 8 March 2019; Accepted: 17 March 2019; Published: 21 March 2019

Abstract: Responsive polymer-based sensors have attracted considerable attention due to their ability to detect the presence of analytes and convert the detected signal into a physical and/or chemical change. High responsiveness, fast response speed, good linearity, strong stability, and small hysteresis are ideal, but to gain these properties at the same time remains challenging. This paper presents a facile and efficient method to improve the photonic sensing properties of polymeric gels by using non-close-packed monolayer colloidal crystals (ncp MCCs) as the template. Poly-(2-vinyl pyridine) (P2VP), a weak electrolyte, was selected to form the pH-responsive gel material, which was deposited onto ncp MCCs obtained by controlled O_2 plasma etching of close-packed (cp) MCCs. The resultant ultrathin photonic polymer gel film (UPPGF) exhibited significant improvement in responsiveness and linearity towards pH sensing compared to those prepared using cp MCCs template, achieving fast visualized monitoring of pH changes with excellent cyclic stability and small hysteresis loop. The responsiveness and linearity were found to depend on the volume and filling fraction of the polymer gel. Based on a simple geometric model, we established that the volume increased first and then decreased with the decrease of template size, but the filling fraction increased all the time, which was verified by microscopy observations. Therefore, the responsiveness and linearity of UPPGF to pH can be improved by simply adjusting the etching time of oxygen plasma. The well-designed UPPGF is reliable for visualized monitoring of analytes and their concentrations, and can easily be combined in sensor arrays for more accurate detection.

Keywords: polymer gel; colloidal crystals; optical film; pH sensor

---

1. Introduction

Developments in society have brought about renewed focus on environmental problems, and a growing need for simple and accurate sensors that can respond to environmental changes has emerged. In nature, there are various creatures that can smartly change color with the environment [1]. Chameleons, for example, can adjust the color of their skin to the color of their surroundings [2]. Zebrafish can change their appearance under light [3]. Some beetles respond to humidity. Inspired by this phenomenon, researchers have synthesized a series of materials that can respond to the surrounding environment by color.

Among various materials, responsive polymer gels are good candidates for mimicking the responsive colors in natures because they are able to change their volume by swelling or deswelling under the stimulation of external conditions and such change can be elaborately converted into optical signals to achieve a visual response to the analyte [4–6]. Based on this mechanism, several colorimetric sensors using polymer gels have been fabricated that respond to pH, temperature, humidity, glucose, macromolecules, and metal ions, among others [7–9].

Photonic crystals, a class of most interesting optical materials, are usually constructed by two or more media with different refractive indices arranged periodically in one, two, or three dimensions (1D, 2D or 3D, respectively) [10]. Because photonic crystals can produce band gaps for photons of a certain frequency, various film colors can be generated. Considering the unique advantages of photonic crystals in optical sensing, researchers have combined them with polymer gels to form advanced response materials. In a pioneer work, Asher's group synthesized a polymer colloid array (PCCA) to quantitatively detect glucose by embedding polymer gel into 3D colloid crystals [11–14]; they also fabricated a 2D hydrogel sensor by attaching a 2D colloid array to hydrogel, and this sensor could change its lattice spacing by swelling of the gel to achieve a visual response to the analyte [15–19]. Polymeric 1D photonic crystals or so-called Bragg stacks have been fabricated by spin-coating of block copolymer solutions [20,21]. Despite their many benefits, however, the problem of slow responses greatly limits the development and application of these sensors. For example, the PCCA glucose sensor requires over 90 min to respond to the analyte [13].

To improve the response speed of sensors, a reverse opal polymer gel was developed using SiO_2 or polystyrene (PS) microspheres as a template [20–28]. The presence of macropores in the film structure enhanced not only the responsiveness of the resulting sensor by increasing its specific surface area but also the transport of the analyte; the resulting response speeds were improved to a certain extent. However, 3D hydrogel pH sensors with an inverse opal structure still require a response time of 20 min to achieve equilibrium [23]. The response speed is proportional to the rate of volume change of polymer gels, and the change rate of gel volume is inversely proportional to the size of the gel [29]. An interfering gel film with a sub-micron size was synthesized by Zhang et al., and the complete response of this film to glucose was achieved within 2 min [30]. Li et al. fabricated reverse opal polymer gel thin films with sub-micron thickness by using MCCs as the template and achieved a fast response to pH within 1 min [31]. However, the stability of these films was weak and the linear relationship between the dip shift of the films and pH could not be maintained under the condition of strong acidity. Recently, we fabricated ultrathin polymer gel films that are infiltrated into MCCs, which show excellent stability in strongly acidic solution [32].

Besides response time, a good linear relationship between the stimulus and the response is of great significance to a sensor. An ideal sensor has not only a simple calibration process but also constant sensitivity and accuracy over the entire measurement range to enable reliable responses to the environment. However, research on the linear relationships of polymeric gel-based sensors is limited. 2D-PCCA films could respond to the analyte by changing the spacing of colloidal crystals, but a poor linear relationship among pH [15], antibiotics [16], and serum was found because the 2D structure of the colloidal array limited lateral swelling to some extent [17]. Although the interfering gel film shows a good linear relationship between the response and glucose concentration, it has very weak optical signals. Thus, the visual response could not be achieved [30].

Hysteresis is another common issue for polymeric gel-based optical sensors. The Donnan potential presents in low-ionic strength solutions and hinders protonation of gels by diffusion and thermodynamic exchange limit elimination of swelling, thus most polymer gel sensors show hysteresis [14]. Serpe's group synthesized a PNIPAm-*co*-AAc microgel that could respond to pH and temperature [33]. The hysteresis loop size of this sensor could be adjusted by using solutions of various ionic strength and changing the concentration of AAc. Although the hysteresis was improved, the phenomenon remained; in theory, however, hysteresis may be completely eliminated after a long period. This problem seriously affects the practical applications of the film for continuous testing.

Targeted at above-mentioned issues, we present herein the fabrication of ultrathin photonic polymer gel films (UPPGF) by using non-close-packed monolayer colloidal crystals (ncp MCCs) as the template. Poly-(2-vinyl pyridine) (P2VP), a weak electrolyte, was chosen for this study because of its pH-dependent swelling ability. Swelling of the polymer gel led to changes in film thickness and shifts in reflection peak. The responsiveness of the proposed sensor was directly related to the volume change of the polymer gel. The filling fraction and total volume of P2VP in the film could be adjusted by controlling the packing density of the ncp MCCs by varying the time of O_2 plasma etching, which was proven to play an important role in improving the responsiveness and linearity of the pH sensor. The film had an overall sub-micron thickness, which promoted ion diffusion and swelling of the gel, leading to fast response speed and small hysteresis loops. The good adhesion between the ncp MCC and the substrate enabled the ordered structure of the film to be maintained, ensuring good cycling stability of the sensor even after repeated testing. The well-designed UPPGF is simple and reliable for visualized monitoring of analytes and their concentrations, and can be easily combined in sensor arrays for more accurate detection by cross-sensing.

2. Materials and Methods

2.1. Raw Material and Reactants

Poly-(2-vinyl pyridine) (P2VP) (M_w: 159 kg mol^{-1}, Fluka), 1,4-diiodobutane (DIB) (99%, Alfa Aesar, UK), nitromethane (NM) (99%, Sinopharm, Shanghai, China), tetrahydrofuran (THF) (99.0%, Guangcheng, Tianjin, China), diethyl ether (DE) (99.5% Fuyu), styrene (95%, Beijing Chemical Co., Beijing, China, washed in NaOH before use), potassium persulfate (99.5%, Beijing Chemical Co., Beijing, China), sodium dodecyl sulfate (SDS) (M_w: 288.38 kg mol^{-1}, Kermel, Tianjin, China), ethanol (99.7%, Sinopharm, Shanghai, China), and ultrapure water ($\geq$18.2 MΩ, Milli-Q Reference, Beijing, China) were used in this work.

2.2. Preparation of Close-Packed Monolayer Colloidal Crystals

First, polystyrene (PS) spheres (438 nm in diameter; standard deviation < 10%) were synthesized by standard emulsion-free polymerization [34]. Si wafers (one side polished) were cut into 1 cm × 1 cm squares, treated with piranha solution, rinsed with copious amounts of water, and dried under a N_2 flow. The cp MCCs were then formed on the silicon wafer by the gas–liquid interface self-assembly method [35]. The adhesive force between the colloidal crystals on the silicon wafers was strengthened by annealing at 80 °C for 24 h. The cp MCCs were etched for different times (2, 4, 6, or 8 min) by O_2 plasma etching with a 150 W power plasma cleaner (Beijing Huiguang Co., Beijing, China) to form ncp MCCs. The oxygen flow rate was maintained at 200 mL/min and 20 °C.

2.3. Preparation of UPPGF

Quaternization of P2VP was carried out following the procedure reported by Tokarev et al. [36]. With some modifications. Briefly, 0.1 g of P2VP and 0.1 mL of DIB were dissolved in a mixture of NM (4 mL) and THF (1 mL) under stirring at room temperature. Then, the solution was heated at 60 °C with stirring for 80 h to accelerate the quarternization reaction between P2VP and DIB. An excess amount of DE was added to the mixture, and centrifugation was performed to eliminate THF and the residual DIB. Then, qP2VP was dissolved in 5 mL of NM to form a 3.25 wt % solution for subsequent spin-coating. The qP2VP solution was spin-coated onto ncp MCCs at a speed of 1000 rpm for 1 min using a spin coater (Laurell-WS650). Finally, UPPGF was obtained after thermal crosslinking at 120 °C for 48 h.

2.4. Characterization

The morphology of the films was examined by a Hitachi SU8010 field emission scanning electron microscope (FE-SEM), and the reflection spectra of the samples were acquired with an Ocean Optics USB2000 fiber optic spectrophotometer coupled to a Leica DM2700 M optical microscope. The reflectance spectra were consistently measured from the same spot of a UPPGF specimen by saving the spot image to identify it in the following experiments. Optical micrographs were taken under white-light LED illumination by a Leica DFC450 digital color camera coupled to a microscope with a 10× objective lens.

2.5. Sensor Test

Solutions of a certain pH were prepared from 0.1 M citric acid (aq.) and 0.1 M trisodium citrate dihydrate (aq.), 0.05 M NaH_2PO_4 (aq.), 0.1 M Na_2HPO_4 (aq.), 0.05 M $NaHCO_3$ (aq.), and 0.1 M NaOH (aq.). The pH of the buffer solution was measured by a pH meter (INESA PHSJ-3F). Each UPPGF sample was dipped into the buffer solution for 2 min and then blown using a N_2 flow to eliminate the excess solution on its surface. The reflectance spectra of the UPPGF samples were recorded before and after the dipping process. The samples were recovered by soaking in pH 10 buffer solution for 2 s, washing with deionized water, and then blowing with N_2.

3. Results and Discussion

3.1. Preparation and Characterization of UPPGF

The synthesis of ultrathin photonic polymer gel films (UPPGF) is shown in Scheme 1. We obtained close-packed (cp) monolayer colloidal crystals (MCCs) by gas–liquid interface self-assembly. As shown in Scheme 1a, the polystyrene (PS) microspheres were stacked in a dense hexagonal manner on the silicon substrate to form cp MCCs. The non-close-packed monolayer colloidal crystals (ncp MCCs-x) (x = etching time in minutes) was obtained by O_2 plasma etching of cp MCCs. P2VP swells by protonation in acid solution.; thus, we selected P2VP as the responsive polymer in this study. The P2VP precursor solution (3.25 wt %, in NM) was immersed into the ncp MCCs by spin-coating (Scheme 1b). Finally, the P2VP was completely cross-linked with DIB in the vacuum drying oven at 120 °C to obtain UPPGF-x (x = etching time in minutes) (Scheme 1c).

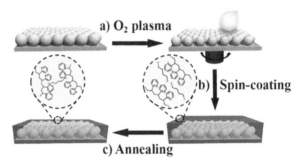

Scheme 1. Steps to fabricate UPPGF. (**a**) Monolayer colloidal crystals of PS nanoparticles were etched through oxygen plasma etching for different times. (**b**) The ncp MCCs was infiltrated with a polymer precursor (qP2VP) solution by spin coating. (**c**) P2VP was crosslinked by annealing.

The packing density of the ncp MCCs can be controlled by adjusting the O_2 plasma etching time. The cp MCCs were treated at 80 °C for 12 h before etching to enhance the contact between PS and the silicon wafer and ensure that the film did not fall off or fold during etching and response detection. O_2 plasma etching was then performed on the cp MCCs. Figure 1 reveals that the particle size of the PS microspheres gradually decreased with increasing etching time (from 438 nm for the sample without etching to 430 nm for ncp MCCs-2, 417 nm for ncp MCCs-4, 404 nm for ncp MCCs-6, and 395 nm for ncp MCCs-8). However, due to the good contact between PS and the silicon wafer, the position of the PS microspheres did not change during the etching process, and the gap between the microspheres increased gradually. Thus, ncp MCCs with different packing density were formed. Low-magnification SEM images reveal that the order of the array was not destroyed by plasma etching, and ordering of

PS microspheres was maintained even after etching for 8 min. Such a characteristic is an important condition enabling films to display color and achieve a visual response to pH.

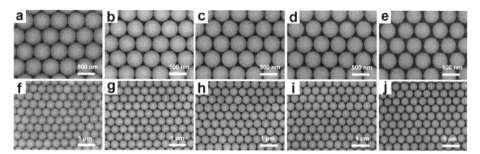

Figure 1. SEM images of the ncp MCCs after etching: (**a,f**) top views of ncp MCCs-0; (**b,g**) top views of ncp MCCs-2; (**c,h**) top views of ncp MCCs-4; (**d,i**) top views of ncp MCCs-6; and (**e,j**) top views of ncp MCCs-8.

UPPGF was prepared by spin-coating a polymer precursor solution onto ncp MCCs and thermal crosslinking. The qP2VP was spin-coated onto the surface of ncp MCCs. As shown in Figure 2, the P2VP was coated uniformly on the surface of the PS microspheres but it did not completely fill the voids between spheres. The viscosity of P2VP is such that rapid spin-coating does not allow it to fully infiltrate the substrate structure. The thickness of the films decreased with increasing etching time (478 nm for qP2VP-infiltrated ncp MCCs-0 to 462 nm for qP2VP-infiltrated ncp MCCs-2, 440 nm for qP2VP-infiltrated ncp MCCs-4, 430 nm for qP2VP-infiltrated ncp MCCs-6, and 414 nm for qP2VP-infiltrated ncp MCCs-8) because the thickness of P2VP on the surface of the PS array was determined by the speed of spin-coating and the concentration of the precursor solution. The thickness of the films depended on the particle size of PS after etching when the speed of the spin-coating and concentration of qP2VP were held constant. Figure 2 shows that the array maintained its good order after thermal cross-linking. Although the temperature of thermal cross-linking was higher than the glass transition temperature of PS, the protective effect of P2VP prevented serious deformation of the microspheres. During thermal crosslinking, P2VP gradually infiltrated the gap between PS microspheres, which grew larger with increasing etching time and allowed more P2VP to infiltrate into the pores. Thus, waves were produced on the surface of the films. Compared with that before thermal cross-linking, the thickness of the films decreased (from 478 to 440 nm for UPPGF-0, from 462 to 419 nm for UPPGF-2, from 440 to 397 nm for UPPGF-4, from 430 to 390 nm for UPPGF-6, and from 414 to 387 nm for UPPGF-8). This finding is related to the slight deformation of PS microspheres and the infiltration of P2VP.

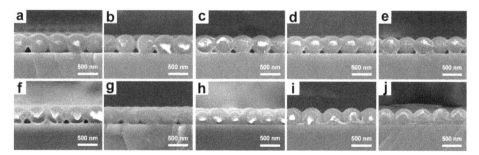

Figure 2. SEM images of the fabricated structures: (**a–e**) cross-sectional views of the qP2VP-infiltrated ncp MCCs-x (x = (**a**) 0 min; (**b**) 2 min; (**c**) 4 min; (**d**) 6 min; and (**e**) 8 min); and (**f–j**) cross-sectional views of UPPGF-x (x = (**f**) 0 min; (**g**) 2 min; (**h**) 4 min; (**i**) 6 min; and (**j**) 8 min) obtained after thermal annealing.

3.2. Optical Properties of UPPGF

To better understand the effect of O_2 plasma etching on the structure of the film, we studied its optical properties. A high-refractive index silicon wafer ($n \sim 3.5$) was chosen as the substrate on which to construct UPPGF. Fabry–Pérot fringes are formed by reflecting the interference between the beams of the thin-film air and thin-film substrate interfaces [37]. Under normal conditions, the position of the interference peak wavelength conforms to Equation (1) [38]:

$$m\lambda = 2nd, \tag{1}$$

where n is the refractive index of the film, m is an integer, and d is the thickness of the film. Calculations indicated that the peak of the cp MCCs was located in the visible region (585 nm; $d = 438$ nm, $m = 2$, and $n = 1.335$) [39]. In the experiments (Figure 3a), the center of the reflection peak was found at 588 nm. The valley observed was the result of multiple scattering from a single sphere, and the characteristic mode of 2D photonic crystals with hexagonal symmetry was found. The position of the valley gradually shifted toward shorter wavelengths with increasing etching time (625 nm for ncp MCCs-2, 617 nm for ncp MCCs-4, 611 nm for ncp MCCs-6, and 596 nm for ncp MCCs-8). This finding could be attributed to the refractive index of the film gradually decreasing with increasing etching time, because the distance between colloidal crystal microspheres did not change with the increase of etching time, but the ratio of air in the array increased, resulting in the decrease of effective refractive index of the film, consistent with the phenomena observed in the SEM images (Figure 1).

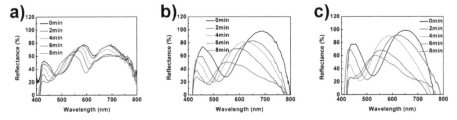

Figure 3. Optical properties of the fabricated structures:(**a**) reflectance spectra of the ncp MCCs-x obtained at different etching times (x = 0, 2, 4, 6, 8 min); (**b**) reflectance spectra of the qP2VP-infiltrated ncp MCCs-x; and (**c**) reflectance spectra of UPPGF-x obtained after thermal annealing.

We constructed UPPGF by spin-coating and thermal crosslinking. We found only one reflection peak in the visible region after spin-coating of the qP2VP, and no valleys associated with the photonic characteristic mode were observed. This is because of the refractive indices contrast was eliminated when the P2VP was infiltrated into the films. In this case, the position of the reflection peak also moved toward shorter wavelengths with increasing etching time (688 nm for qP2VP-infiltrated ncp MCCs-0, 653 nm for qP2VP-infiltrated ncp MCCs-2, 619 nm for qP2VP-infiltrated ncp MCCs-4, 584 nm for qP2VP-infiltrated ncp MCCs-6, and 549 nm for qP2VP-infiltrated ncp MCCs-8; Figure 3b). Because the thickness of the film decreased gradually, the UPPGF obtained by thermal cross-linking showed the same trend (656 nm for UPPGF-0, 621 nm for UPPGF-2, 589 nm for UPPGF-4, 547 nm for UPPGF-6, and 533 nm for UPPGF-8; Figure 3c). The reflection peaks of all films demonstrated a certain blue-shift after thermal cross-linking, which was due to the decrease in film thickness. The above data are consistent with the change in film thickness observed in the SEM images (Figure 2).

3.3. Responsiveness of the UPPGF to pH and Mechanism Research

We tested the responsiveness of the UPPGF sensors to pH by immersing them in buffers of different pH. P2VP swells in acidic solution, and its degree of swelling is related to its degree of protonation. Figure 4 illustrates that the reflection peaks of the films did not change significantly when the films were immersed in alkaline solution (pH ≥ 7). In acidic solution (pH < 7), however, the reflection peaks of all films gradually shifted toward longer wavelengths with decreasing pH. Even under strongly acidic (pH = 2.57) conditions, this response was maintained because the PS array prevented the collapse of the film structure caused by the high degree of swelling. This phenomenon is shown more intuitively in Figure S5. As shown in Figure 4f and Figure S5, in comparison with that of UPPGF-0, the pH responsiveness of the UPPGF templated by non-close-packed monolayer colloidal crystals was improved, and UPPGF-4 and UPPGF-6 showed the best responsiveness to pH. The displacement of reflection peak of UPPGF-4 was 80 nm, about 30 nm longer than the wavelength shift of UPPGF-0. However, compared with that of UPPGF-6, the responsiveness of UPPGF-8 was reduced to a certain extent. It is well known that the sensing ability of polymer gel sensor is closely related to the volume of polymer gel. Thus, we think that the responsiveness of the UPPGF was determined by the volume of P2VP, and the volume change of P2VP in the UPPGF may be influenced by etching.

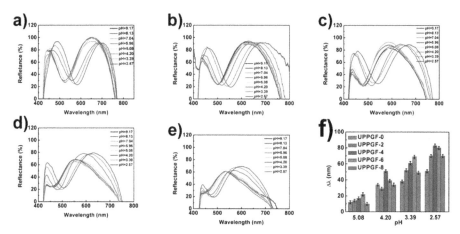

Figure 4. (**a–e**) pH dependence of the reflectance spectra of the UPPGF-x (x = (**a**) 0 min; (**b**) 2 min; (**c**) 4 min; (**d**) 6 min; (**e**) 8 min) sensors after equilibration in pH buffer solution; and (**f**) the corresponding shifts of the reflectance peak of UPPGF-x.

We further explained the variation of the pH responsiveness of the UPPGF with increasing etching time through simple calculation based on geometric model. In Scheme 2, we set the diameter of the PS microspheres as D and the total thickness of the film as H. The D of the PS microspheres decreased to d after etching, assuming that the thickness of P2VP on the PS microspheres h is unchanged. We then calculated the volume (V_{P2VP}) of P2VP with decreasing d:

$$V_{P2VP} = V_{total} - V_{PS} = \frac{3\sqrt{3}D^2H - \pi d^3}{12} = \frac{3\sqrt{3}D^2(h+d) - \pi d^3}{12} \tag{2}$$

$$V'_{P2VP} = \frac{\sqrt{3}D^2 - \pi d^2}{4} \tag{3}$$

$$\Delta\lambda = \frac{2n \cdot \Delta H}{m} = \frac{2n \cdot V_{P2VP} \cdot f_{P2VP}}{m} \tag{4}$$

Polymers **2019**, *11*, 534

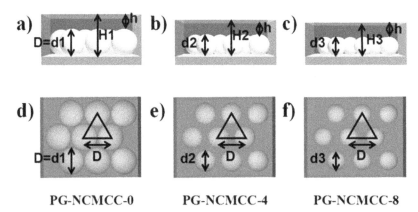

PG-NCMCC-0 **PG-NCMCC-4** **PG-NCMCC-8**

Scheme 2. (**a–e**) Side views of UPPGF-0, UPPGF-4, and UPPGF-8; and (**d–f**) top views of UPPGF-0, UPPGF-4, and UPPGF-8.

We found that the V_{P2VP} did not always increase with decreasing d. Using Equations (2) and (3), we calculated that V_{P2VP} increased gradually as D decreased from D to $0.74D$ but decreased gradually with further decreases in particle size beyond $0.74D$, i.e., V_{P2VP} reached its maximum value at $0.74D$. According to Equation (4) (where $\Delta\lambda$ is the displacement of the reflection peak, ΔH is the change in thickness of the film, n is the refractive index of the film, m is an integer, and f_{P2VP} is the swelling rate of P2VP), the displacement of the reflection peak is proportional to the variation in V_{P2VP}. Therefore, the law of responsiveness of UPPGF to pH is well explained. In the actual tests, however, maximum volume was achieved even if the particle size was not reduced to $0.74D$, likely because P2VP did not completely cover the film, as assumed, to form a smooth surface (Figure 2). Thus, V_{P2VP} in the actual experiments decreased at a faster rate than predicted by the calculations.

We monitored the change in film thickness with decreasing pH through SEM to confirm our hypothesis. In Figure 5, the thickness of UPPGF-4 increased from 397 to 408, 419, 440, and 449 nm in response to immersion in pH 5.08, 4.20, 3.39, and 2.57 buffer solutions, respectively; the thickness of the UPPGF would still increase when pH is 2.57 with no structural collapse. Other films showed the same trend with decreasing pH (Figures S1–S4). UPPGF-4 also showed the maximum thickness variation, which explains why UPPGF-4 had the best pH responsiveness.

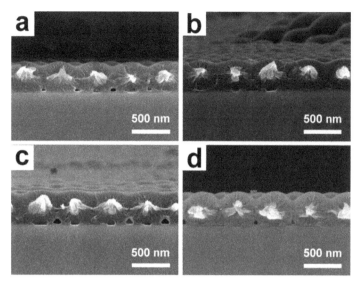

Figure 5. SEM images of UPPGF-4 after equilibration in buffer solution at: pH 5.08 (**a**); pH 4.20 (**b**); pH 3.39 (**c**); and pH 2.57 (**d**).

3.4. Linearity of the UPPGF to pH and Mechanism Research

Sensors with good linearity can detect analytes more accurately in the whole range of measurement. As shown in Figure 6a, we found that the linearity towards pH was gradually enhanced with increasing etching time. The coefficient of determination (R^2) of the sensor gradually increased from 0.95684 (UPPGF-0) to 0.9816 (UPPGF-8). This phenomenon can also be explained by Scheme 2. The relationship between the filling fraction of P2VP (N_{P2VP}) and PS particle size (d) is in accordance with Equation (5):

$$N_{P2VP} = 1 - \frac{\pi d^3}{3\sqrt{3}D^2(h+d)} \tag{5}$$

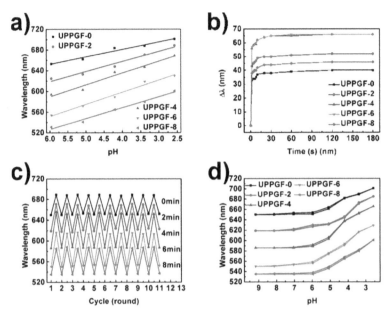

Figure 6. (a) Linear relation between UPPGF-x (x = 0, 2, 4, 6, 8 min) and pH; (b) response kinetics of UPPGF-x to a pH 3.39 buffer solution; (c) reversibility of UPPGF-x over 10 cycles of exposure to pH 9.17–3.39; and (d) hysteresis loops of the UPPGF-x sensors between pH 9.17 and 2.57.

Equation (5) reveals that N_{P2VP} always increase with decreasing d. We thus considered that N_{P2VP} is an important factor affecting the linearity of response. Although the PS array provided the necessary optical signals for the film, it could restrain the regular swelling of the polymer gel. Therefore, the linearity of UPPGF increased with increasing etching time. Considering response degrees and linearity of response, UPPGF-6 exhibited the best properties among the synthesized films.

3.5. Response Speed of the UPPGF to pH

Fast response speeds are widely favored in practical applications. We measured the response speed of the UPPGF by immersing them in buffer solution of pH = 3.39 and recording the change in reflection peak over time. Approximately 90% of the total response could be achieved within 10 s by the films, and stable responses could be achieved within 2 min (Figure 6b). The response speed of UPPGF was faster than that of the inverse opal hydrogel pH sensor (20 min) reported by Braun et al. [23] and the 2D-PCCA pH sensor (30 min) reported by Asher et al. [15], and comparable with our previously reported inverse opal monolayers of P2VP gels [31] and ultrathin P2VP gel-infiltrated MCCs films [32]. This fast response speed was due to the structural characteristics of the UPPGF, which included submicron thickness. The response speeds of UPPGF did not decrease with increasing volume of P2VP. Although the volume of P2VP increased after etching, the thickness of the whole film decreased.

3.6. Stability of the UPPGF to pH

We tested the stability of the UPPGF samples. Folding or shedding was not found in the SEM images when the UPPGF responds to the pH buffer solution (Figure 5 and Figures S1–S4). Then, the films were immersed in buffer solution of pH = 3.39 and then recovered in a solution of pH = 9.17. Figure 6c reveals that the position of the reflection peak remains basically unchanged after 10 cycles, thereby indicating good cyclic stability. Although more P2VP was in contact with the substrate with increasing etching time, the contact force between PS and the silicon substrate was improved through heat treatment of the cp MCCs at 80 °C prior to etching. Thus, the UPPGF sensors showed good stability.

3.7. Hysteresis Loops of the UPPGF to pH

Sensors with low hysteresis loops can accurately detect the environment regardless of their input history. The peak shifts of the UPPGF samples observed under two approaches of pH input, i.e., from pH 9 to 2 and from pH 2 to 9, were recorded, and film recovery was found to be unnecessary for detecting different pH solution. As shown in Figure 6d, all UPPGF exhibited small hysteresis loops with short test times (2 min), likely because of the submicron thickness of the film. Ions could diffuse rapidly through the film, and the polymer gel could swell and shrink quickly. Thus, UPPGF can be used to test the environment continuously.

3.8. Visual Response of the UPPGF to pH

Importantly, the UPPGF achieved visual response to pH. In Figure 7, we can see that the color of the film changed with the change of pH. Compared with UPPGF-0, the color change of the UPPGF templated by ncp MCCs was more obvious. This phenomenon is consistent with the change of the redshift of the reflection peak with the etching time (Figure 4). Moreover, other responsive materials (for example, other responsive polymer gels, MOF, etc.) can be combined with ncp MCCs to improve their sensing capabilities. A sensor array could be obtained by the combination of the films etched at different times to achieve more accurate analysis.

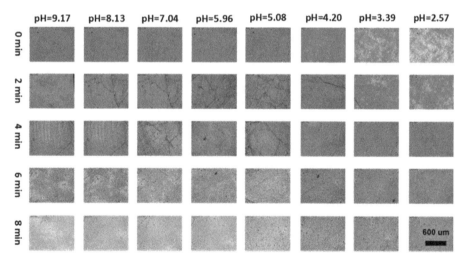

Figure 7. Optical micrographs showing the pH dependence of the different response colors of UPPGF-x (x = 0, 2, 4, 6, 8 min) (the size of each optical micrograph: 1.2 mm × 0.9 mm).

4. Conclusions

In summary, UPPGF using non-close-packed monolayer colloidal crystals (ncp MCCs) as template were prepared by spin-coating of qP2VP and subsequent thermal cross-linking. The UPPGF showed bright structural colors depending on the thickness of the film. The swelling of P2VP gel in acidic solution caused changes in the film thickness, thus resulting in a change in the visible color. Compared with those templated by cp MCCs, the UPPGF not only increased the wavelength shift of the reflection peak upon pH changes, but also improved the linearity of response, while maintaining the advantages of fast response speed, high cycle stability and small hysteresis loop. The packing density of ncp MCCs could be adjusted by controlling the time of oxygen plasma etching to effectively regulating the volume and filling factor of P2VP, which could be used to improve the responsiveness and linearity. The present strategy of fine-tuning the volume of filling fraction of polymeric gel in a sensor device can be feasibly extended to other responsive materials/systems to improve the responsiveness and linearity, as well as makes it simple to construct sensor arrays to enable more accurate detections.

Supplementary Materials: The following are available online at http://www.mdpi.com/2073-4360/11/3/534/s1, Figure S1: SEM images of UPPGF-0 after equilibration in pH 5.08, pH 4.20, pH 3.39, and pH 2.57 buffer solution; Figure S2: SEM images of UPPGF-2 after equilibration in pH 5.08, pH 4.20, pH 3.39, and pH 2.57 buffer solution; Figure S3: SEM images of UPPGF-6 after equilibration in pH 5.08, pH 4.20, pH 3.39, and pH 2.57 buffer solution; Figure S4: SEM images of UPPGF-8 after equilibration in pH 5.08, pH 4.20, pH 3.39, and pH 2.57 buffer solution; Figure S5: Corresponding shifts of the reflectance peak of UPPGF-x (x = 0 min, 2 min, 4 min, 6 min, 8 min) after equilibration in pH buffer solution.

Author Contributions: Conceptualization, C.L.; methodology, C.L. and S.Y.; software, S.Y. and S.D.; validation, C.L., S.Y., D.C. and X.J.; formal analysis, C.L. and D.C.; investigation, C.L. and S.Y.; resources, D.C. and X.J.; data curation, S.Y. and C.L.; writing—original draft preparation, C.L. and S.Y.; writing—review and editing, C.L. and S.Y.; visualization, C.L., S.Y. and S.D.; supervision, C.L., D.C. and X.J.; and project administration, C.L., D.C. and X.J.

Funding: This research received no external funding.

Acknowledgments: The authors acknowledge support from the National Natural Science Foundation of China (NSFC 21771118), and the Taishan Scholars Climbing Program of Shandong Province (Grant tspd20150201).

Conflicts of Interest: The authors declare no conflict of interest.

References

1. Isapour, G.; Lattuada, M. Bioinspired Stimuli-Responsive Color-Changing Systems. *Adv. Mater.* **2018**, *30*, 1707069. [CrossRef]

2. Teyssier, J.; Saenko, S.V.; van der Marel, D.; Milinkovitch, M.C. Photonic crystals cause active colour change in chameleons. *Nat. Commun.* **2015**, *6*, 6368. [CrossRef] [PubMed]

3. Rossiter, J.; Yap, B.; Conn, A. Biomimetic chromatophores for camouflage and soft active surfaces. *Bioinspir. Biomim.* **2012**, *7*, 036009. [CrossRef] [PubMed]

4. Takeoka, Y.; Watanabe, M. Tuning Structural Color Changes of Porous Thermosensitive Gels through Quantitative Adjustment of the Cross-Linker in Pre-gel Solutions. *Langmuir* **2003**, *19*, 9104–9106. [CrossRef]

5. Stumpel, J.E.; Broer, D.J.; Schenning, A.P.H.J. Stimuli-responsive photonic polymer coatings. *Chem. Commun.* **2014**, *50*, 15839–15848. [CrossRef]

6. Lee, J.H.; Koh, C.Y.; Singer, J.P.; Jeon, S.J.; Maloovan, M.; Stein, O.; Thomas, E.L. 25th anniversary article: Ordered polymer structures for the engineering of photons and phonons. *Adv. Mater.* **2014**, *26*, 532–569. [CrossRef] [PubMed]

7. Stuart, M.A.C.; Huck, W.T.S.; Genzer, J.; Marcus, M.; Ober, C.; Stamm, M.; Sukhorukov, G.B.; Szleifer, I.; Tsukruk, V.V.; Urban, M.; et al. Emerging applications of stimuli-responsive polymer materials. *Nat. Mater.* **2010**, *9*, 101–103. [CrossRef]

8. Hendrickson, R.G.H.; Lyon, L.A. Bioresponsive hydrogels for sensing applications. *Soft Matter* **2009**, *5*, 29–35. [CrossRef]

9. Wei, M.; Gao, Y.F.; Li, X.; Serpe, M.J. Stimuli-responsive polymers and their applications. *Polym. Chem.* **2017**, *8*, 127–143. [CrossRef]

10. Paquet, C.; Kumacheva, E. Nanostructured polymers for photonics. *Mater. Today* **2008**, *11*, 48–56. [CrossRef]

11. Holtz, J.H.; Asher, S.A. Polymerized colloidal crystal hydrogel films as intelligent chemical sensing materials. *Nature* **1997**, *389*, 829–832. [CrossRef] [PubMed]

12. Asher, S.A.; Alexeev, V.L.; Goponenko, A.V.; Sharma, A.C.; Lednev, I.K.; Wilcox, C.S.; Finegold, D.N. Photonic crystal carbohydrate sensors: Low ionic strength sugar sensing. *J. Am. Chem. Soc.* **2003**, *125*, 3322–3329. [CrossRef] [PubMed]

13. Ben-moshe, M.; Alexeev, V.L.; Asher, S.A. Fast responsive crystalline colloidal array photonic crystal glucose sensors. *Anal. Chem.* **2006**, *78*, 5149–5157. [CrossRef] [PubMed]

14. Xu, X.L.; Goponenko, A.V.; Asher, S.A. Polymerized polyHEMA photonic crystals: pH and ethanol sensor materials. *J. Am. Chem. Soc.* **2008**, *130*, 3113–3119. [CrossRef] [PubMed]

15. Zhang, J.T.; Wang, L.L.; Luo, J.; Tikhonov, A.; Kornienko, N.; Asher, S.A. Photonic crystal chemical sensors: pH and ionic strength. *J. Am. Chem. Soc.* **2011**, *133*, 9152–9155. [CrossRef] [PubMed]
16. Zhang, J.T.; Chao, X.; Liu, X.L.; Asher, S.A. Two-dimensional array Debye ring diffraction protein recognition sensing. *Chem. Commun.* **2013**, *49*, 6337–6339. [CrossRef] [PubMed]
17. Cai, Z.Y.; Zhang, J.T.; Xue, F.; Hong, Z.M.; Punihaole, D.; Asher, S.A. 2D photonic crystal protein hydrogel coulometer for sensing serum albumin ligand binding. *Anal. Chem.* **2014**, *86*, 4840–4847. [CrossRef]
18. Cai, Z.Y.; Smith, N.L.; Zhang, J.T.; Asher, S.A. Two-dimensional photonic crystal chemical and biomolecular sensors. *Anal. Chem.* **2015**, *87*, 5013–5025. [CrossRef] [PubMed]
19. Cai, Z.Y.; Luck, L.A.; Punihaole, D.; Madurac, J.D.; Asher, S.A. Photonic crystal protein hydrogel sensor materials enabled by conformationally induced volume phase transition. *Chem. Sci.* **2016**, *7*, 4557–4562. [CrossRef]
20. Kang, H.S.; Lee, J.; Cho, S.M.; Park, T.H.; Kim, M.J.; Park, C.; Lee, S.W.; Kim, K.L.; Ryu, D.Y.; Huh, J. Printable and rewritable full block copolymer structural color. *Adv. Mater.* **2017**, *29*, 1700084. [CrossRef]
21. Park, T.H.; Yu, S.; Cho, S.H.; Kang, H.S.; Kim, Y.; Kim, M.J.; Eoh, H.; Park, C.; Jeong, B.; Lee, S.W.; et al. Block copolymer structural color strain sensor. *NPG Asia Mater.* **2018**, *10*, 328–339. [CrossRef]
22. Cassagneau, T.; Caruso, F. Conjugated polymer inverse opals for potentiometric biosensing. *Adv. Mater.* **2002**, *14*, 1837–1841. [CrossRef]
23. Lee, Y.J.; Braun, P.V. Tunable inverse opal hydrogel pH sensors. *Adv. Mater.* **2003**, *15*, 563–566. [CrossRef]
24. Xu, X.B.; An, Q.; Li, G.T.; Tao, S.Y.; Liu, J. Imprinted photonic polymers for chiral recognition. *Angew. Chem. Int. Ed.* **2006**, *45*, 8145–8148.
25. Hong, W.; Hu, X.B.; Zhao, B.Y.; Zhang, F.; Zhang, D. Tunable photonic polyelectrolyte colorimetric sensing for anions, cations and zwitterions. *Adv. Mater.* **2010**, *22*, 5043–5047. [CrossRef] [PubMed]
26. Zhang, C.J.; Cano, G.G.; Braun, P.V. Linear and fast hydrogel glucose sensor materials enabled by volume resetting agents. *Adv. Mater.* **2014**, *26*, 5678–5683. [CrossRef] [PubMed]
27. Couturier, J.P.; Sutterlin, M.; Laschewsky, A.; Hettrich, C.; Wischerhoff, E. Responsive inverse opal hydrogels for the sensing of macromolecules. *Angew. Chem. Int. Ed.* **2015**, *54*, 6641–6644. [CrossRef] [PubMed]
28. Shin, J.; Han, S.G.; Lee, W. Dually tunable inverse opal hydrogel colorimetric sensor with fast and reversible color changes. *Sens. Actuators B* **2012**, *168*, 20–26. [CrossRef]
29. Tanaka, T.; Fillmore, D.J. Kinetics of swelling of gels. *J. Chem. Phys.* **1979**, *70*, 1214–1218. [CrossRef]
30. Zhang, X.; Guan, Y.; Zhang, Y.J. Ultrathin hydrogel films for rapid optical biosensing. *Biomacromolecules* **2012**, *13*, 92–97. [CrossRef]
31. Li, C.; Lotsch, B.V. Stimuli-responsive 2D polyelectrolyte photonic crystals for optically encoded pH sensing. *Chem. Commun.* **2012**, *48*, 6169–6171. [CrossRef]
32. Yu, S.M.; Han, Z.M.; Jiao, X.L.; Chen, D.R.; Li, C. Ultrathin polymer gel-infiltrated monolayer colloidal crystal films for rapid colorimetric chemical sensing. *RSC Adv.* **2016**, *6*, 66191–66196. [CrossRef]
33. Hu, L.; Serpe, M.J. Controlling the response of color tunable poly (N-isopropylacrylamide) microgel-based etalons with hysteresis. *Chem. Commun.* **2013**, *49*, 2649–2651. [CrossRef]
34. Holland, B.T.; Blanford, C.F.; Do, T.; Stein, A. Synthesis of highly ordered, three-dimensional, macroporous structures of amorphous or crystalline inorganic oxides, phosphates, and hybrid composites. *Chem. Mater.* **1999**, *11*, 795–805. [CrossRef]
35. Li, C.; Hong, G.; Wang, P.; Yu, D.; Qi, L. Wet chemical approaches to patterned arrays of well-aligned ZnO nanopillars assisted by monolayer colloidal crystals. *Chem. Mater.* **2009**, *21*, 891–897. [CrossRef]
36. Tokarev, I.; Orlov, M.; Minko, S. Responsive polyelectrolyte gel membranes. *Adv. Mater.* **2006**, *18*, 2458–2460. [CrossRef]
37. Romanov, S.G.; Vogel, N.; Bley, K.; Landfester, K.; Weiss, C.K.; Orlov, S.; Korovin, A.V.; Chuiko, G.P.; Regensourger, A.; Romanova, A.S.; et al. Probing guided modes in a monolayer colloidal crystal on a flat metal film. *Phys. Rev. B* **2012**, *86*, 195145. [CrossRef]

38. Lu, G.; Hupp, J.T. Metal-Organic Frameworks as Sensors: A ZIF-8 Based Fabry-Pérot Device as a Selective Sensor for Chemical Vapors and Gases. *J. Am. Chem. Soc.* **2010**, *132*, 7832–7833. [CrossRef]

39. López-Garcia, M.; Galisteo-Lóper, J.F.; López, C.; Garcia-Martín, A. Light confinement by two-dimensional arrays of dielectric spheres. *Phys. Rev. B* **2012**, *85*, 235145. [CrossRef]

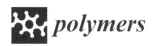

Article

Hardware and Software Development for Isotonic Strain and Isometric Stress Measurements of Linear Ionic Actuators

Madis Harjo [1], Tarmo Tamm [1], Gholamreza Anbarjafari [2] and Rudolf Kiefer [3,*]

[1] Institute of Technology, University of Tartu, Nooruse 1, 50411 Tartu, Estonia; madis.harjo@gmail.com (M.H.); tarmo.tamm@ut.ee (T.T.)

[2] iCV Research Lab, Institute of Technology, University of Tartu, Tartu 50411, Estonia; shb@ut.ee

[3] Conducting polymers in composites and applications Research Group, Faculty of Applied Sciences, Ton Duc Thang University, Ho Chi Minh City 850000, Vietnam

* Correspondence: rudolf.kiefer@tdtu.edu.vn

Received: 18 April 2019; Accepted: 16 June 2019; Published: 17 June 2019

Abstract: An inseparable part of ionic actuator characterization is a set of adequate measurement devices. Due to significant limitations of available commercial systems, in-house setups are often employed. The main objective of this work was to develop a software solution for running isotonic and isometric experiments on a hardware setup consisting of a potentiostat, a linear displacement actuator, a force sensor, and a voltmeter for measuring the force signal. A set of functions, hardware drivers, and measurement automation algorithms were developed in the National Instruments LabVIEW 2015 system. The result is a software called isotonic (displacement) and isometric (force) electro-chemo-measurement software (IIECMS), that enables the user to control isotonic and isometric experiments over a single compact graphical user interface. The linear ionic actuators chosen as sample systems included different materials with different force and displacement characteristics, namely free-standing polypyrrole films doped with dodecylbenzene sulfonate (PPy/DBS) and multiwall carbon nanotube/carbide-derived carbon (MWCNT-CDC) fibers. The developed software was thoroughly tested with numerous test samples of linear ionic actuators, meaning over 200 h of experimenting time where over 90% of the time the software handled the experiment process autonomously. The uncertainty of isotonic measurements was estimated to be 0.6 μm (0.06%). With the integrated correction algorithms, samples with as low as 0 dB signal-to-noise ratio (SNR) can be adequately described.

Keywords: IIECMS; MWCNT-CDC fibers; PPy/DBS linear films; uncertainty measurements

1. Introduction

Ionic actuators are an interesting class of materials which can change their shape over electrochemical stimuli [1]. "Ionic" means that these materials rely on ions (from electrolyte) to carry out their shape change. These materials are also called "artificial muscles" [2] due to relative similarity to the behavior of natural muscles. Applications can be found in soft robotics [3,4], smart textiles [5], energy harvesters [6,7], and biomedical applications [8]. From the variety of materials belonging to this class, conducting polymers [9] and multiwall carbon nanotube/carbide-derived carbon (MWCNT-CDC) fibers [10] were chosen, which differ in their actuation mechanism and formation. Conducting polymers, such as polypyrrole, are a class of materials that follow redox-active processes (Faradaic actuators) [11]. During oxidation a positive charge is formed on the PPy network, in order to maintain the electroneutrality, counterions (anions with solvent) move into the PPy film [12]. Consequently, the polymer film swells (shape change), and during reduction those anions leave the film (anion-driven actuator). The whole process is reversible. Alternatively, if the anions are too large or too strongly interacting, like dodecylbenzene sulfonate (DBS), they become immobile inside the PPy film, while during reduction the charge of DBS⁻ anions is compensated by the ingress of cations into the film (cation-driven actuators, like PPy/DBS [13]). MWCNT-CDC fibers are formed by dielectrophoresis [14], where the fiber is drawn from a droplet of mixed MWCNT and CDC solution over a sharp needle under AC voltage. The MWCNT-CDC actuation mechanism follows the electrical double layer (EDL) processes (non-Faradaic actuators [15]), where the charging of the double layer results in C-C bond length change [16], but the actuation is mostly explained by ion migration [17,18].

For the development and characterization of actuator materials, an accurate measurement setup is an important tool. There have been several hardware constructions using different sensors, for example isometric transducers to measure force, and isotonic transducers to measure strain. For each sensor system, an individual experimental setup is required [19,20]. Recently, a unique hardware setup based on linear variable distance transducers [21] was reported for the measurement of the length change (strain) of conducting polymer devices. The setup consisted of a beam balance with constant weight on one side and the PPy helix tube on the other side. The disadvantage of this setup is the absence of any isometric (force) measurements. Some other constructions have been proposed based on optical signals (lasers) [22], whereas on the end of the hung film a thin wire suspends a reflecting plate through a pinhole of the cell and the up and down movement can be measured by laser displacement. In general, the combination of an isometric transducer with a linear actuation stage (LAS) has been shown to be capable of strain and stress measurements without changing the setup [23–25]. The main problem with all these hardware setups is the neglect of synchronization between the driving potentiostat and the linear actuation device.

Our main interest, therefore, was to focus on a suitable software solution to perform synchronized isotonic and isometric measurements at the same time scale. Here, we want to demonstrate the suitability of the software solution on two different materials: one having high strain and stress—the conducting polymer (PPy/DBS), the other pushing the limitations of the software solution with low strain—the MWCNT-CDC fiber. Numerous manufacturers of stress and strain measurement equipment have advertised various software solutions, which provide isotonic or isometric measurement functionalities with driving signal generation possibilities [26–28]. The primary drawback of these systems, which in general are composed of two different devices (a potentiostat, a linear muscle analyzer with force sensor), is that the devices start and run independently from each other, [29] leading to disconnected time scales/time frames of the measurements— meaning poor synchronizability. Aurora Scientific [27] does provide a complete set of hardware (isotonic/isometric sensor, signal generator, data acquisition device) with a LabVIEW-based software (named dynamic measurement control (DMC)) that controls the hardware resources, executes experiments, and analyzes measured data [27,30,31]. Despite the apparent suitability of the DMC software and other similar products, these solutions are rarely available as open source. Even if they were, the isotonic/isometric measurement software solutions are directly designed for a specific hardware setup [28,32]. Another drawback is the amount of data, collected in binary format and in MB sizes, as the lowest sampling rates available are in the order of 1000 points/s, which make the data analysis rather complicated. Therefore, our interest was to obtain a flexible isotonic and isometric measurement solution that can be composed of devices from different manufacturers and handle data in a manageable text file format in kB scale. The most resource-efficient solution was to develop a custom software for running the desired experiments meeting the requirements.

Here, we present the resulting measurement system and demonstrate its performance on two model ionic actuator systems—a conducting polymer film and a carbon composite fiber.

2. Material and Methods

2.1. Materials

Solvents: propylene carbonate (PC, 99%) and ethylene glycol (EG, 99.8%) were purchased from Fluka (Bucharest, Romania). Pyrrole (Py, ≥ 98%, Sigma Aldrich, Taufkirchen, Germany) was vacuum-distilled prior to use and stored at a low temperature in the dark. Sodium dodecylbenzene-sulfonate (NaDBS, technical grade) and bis(trifluoromethane) sulfonamide lithium salt (LiTFSI, 99%), polyvinyl pyrrolidone (PVP, average mol. wt. 40,000) were acquired from Sigma Aldrich (Taufkirchen, Germany). MilliQ+ water was used for aqueous solutions. Amorphous titanium carbide (TiC)-derived carbon (CDC TiC-800) was purchased from Skeleton Technologies Ltd (Großröhrsdorf, Germany). Multiwall carbon nanotubes (MWCNT) were purchased from Sigma Aldrich(Taufkirchen, Germany) (O.D. × I.D. × L = 7–15 nm × 3–6 nm × 0.5–200 μm).

2.2. Formation of Ionic Electroactive Materials

2.2.1. Multiwall Carbon Nanotube/Carbide-Derived Carbon (MWCNT-CDC) Fiber

The MWCNT and CDC material were dispersed in deionized water containing PVP as the surfactant and sonicated with a UPS200S (Hielscher, Teltow, Germany) ultrasonic processor for 30 min at 50% amplitude. PVP, CDC, MWCNT, and water were mixed at a ratio of 1:2 and: 4:1500 (wt %), respectively. The surfactant was added to stabilize the dispersion for extended periods. Fibers were prepared according to the method described recently [14]. Briefly, the tip of a chemically etched (diameter 0.1 μm) tungsten needle was inserted into a droplet of MWCNT-CDC suspended in deionized water deposited onto a stainless-steel plate. Then, an AC voltage was applied between the tungsten tip and the metal plate (0–350 Vpp, 2 MHz), and the tungsten tip was withdrawn until a MWCNT-CDC fiber of the desired length was formed. The initial spacing between the needle and the plate electrode (i.e., the height of the droplet) was 3 mm. The retraction of the needle was performed with an M-413.3PD precision stage (Physik Instrumente, Karlsruhe, Germany). MWCNT-CDC fibers with embedded CDC particles in the range of 25% were obtained (with an average diameter of 150 μm) and dried in an oven at 100 °C.

2.2.2. Polypyrrole Doped with Dodecylbenzene-Sulfonate (PPy/DBS)

PPy/DBS was polymerized galvanostatically at 0.1 mA cm^{-2} (40,000s) in a 2-electrode cell with a stainless-steel mesh counter electrode and a stainless-steel sheet working electrode (18 cm^2) in a 0.1 M NaDBS, 0.1 M Py, in EG/Milli-Q (1:1) mixture. The temperature of the polymerization was −40 °C. The obtained PPy/DBS films were washed in ethanol to remove excess of pyrrole with additional washing steps in MilliQ+ water to remove excess of NaDBS, and dried in an oven. The film, in thickness of 18.5 μm, were then stored in LiTFSI propylene carbonate (0.2 M) electrolyte.

2.3. Linear Muscle Analyzer Set up

To measure strain and stress without changing the test setup, a Linear Actuation Staging (LAS) with an isometric transducer (force sensor) were combined. The setup is shown in Figure 1a and the connection hierarchy in Figure 1b.

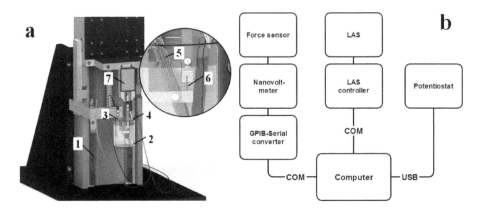

Figure 1. a: Stress and strain measurement (linear muscle analyzer) setup [10]. (1) Linear Actuation Staging (LAS), (2) beaker with electrolyte, (3)(4)(5) electrodes of the potentiostat, (6) ionic electroactive material sample, (7) force sensor. **b**: Connection hierarchy of the devices used for measurement setup.

Figure 1a shows that the sample (6) is attached between two clamps. The lower clamp is static and is fixed to the chassis (1) of the LAS (Linear actuation stage). The upper clamp is attached to a force sensor (7) which is mounted on a plate of the LAS (Physik instrumente M-414.3PD, min step size 0.5 μm, Karlsruhe, Germany) that enables it to perform linear movements. Figure 1b shows the hierarchy of the devices. The signal of the force sensor (Panlab TRI202PAD, Barcelona, Spain) is read by a voltmeter (Keithley 2182A, Beaverton, United States) which exchanges data with the computer over a General-Purpose Interface Bus (GPIB) to Universal Serial Bus (USB) converter (Prologix Rev 6.4.1, Asheville, North Carolina). The LAS (Physik Instrumente M-414.3PD, Karlsruhe, Germany) is controlled by a controller (Physik Instrumente C-863, Karlsruhe, Germany), which accepts commands from the computer via a serial port. Lastly, the potentiostat (Biologic PG581, Göttingen, Germany) is connected to the computer using a USB interface.

To examine the samples, PPy/DBS films and MWCNT-CDC fibers were cut in strips of 1.0 cm * 0.1 cm and fixed between the force sensor and on the fixed arm with gold contacts that served as a working electrode in the linear muscle analyzer setup (Figure 1). The initial length of the films between the clamps was 1 mm. The strain, ε, was calculated from the formula $\varepsilon = \Delta l/l$, where Δl refers to $l - l_1$ with l as the original length of the film (1 mm) and l_1 the change of length obtained from isotonic measurements. The load applied on PPy/DBS films was 6 g (58.8 mN, 3.2 MPa) and the load on MWCNT-CDC films was 60 mg (0.6 mN, 33.2 kPa). For isometric (force) measurements, the stress, σ, was calculated using the formula $\sigma = F/A$, where F is the force in N acting on an object ($F = m^*g$ with m the mass and g means acceleration due to gravity as a constant 9.8 m s^{-2}) and A represents the cross-sectional area of the object (width * thickness). A platinum sheet was used as the counter electrode in the measurements cell and Ag/AgCl (3M KCl) as the reference electrode. For PPy/DBS films and for MWCNT-CDC fibers, 0.2 M LiTFSI propylene carbonate solution was applied. For isotonic and isometric measurements, cyclic voltammetry (scan rate 5 mV s^{-1}) was used for driving in the potential range of 0.65 to −0.6 V.

Characterization of the Materials

PPy/DBS films and MWCNT-CDC fibers were examined by scanning electron microscopy (Helios NanoLab 600, FEI, Hillsboro, OR, USA). The electrical conductivity of the fibers was measured by a simple two-point probe method.

2.4. Linear Muscle Analyzer Software Description

The software of the measurement system was developed [33] in the LabVIEW 2015 environment, and it provides different control modes (Figure S1), which allows it to calibrate the force sensor, test sample elasticity, run electrochemical analysis with the potentiostat (real time measurements), and measure weight change (stress) and length change (strain). For real time measurement requirements, the shortest data acquisition period must be 1s, and after the user has activated an experiment, the devices must be initialized in 10s.

2.4.1. Calibration of the Force Sensor

Calibration of the force sensor is necessary for converting the force sensor's output voltage into values, which represent milligrams. To operate the software, the measurement called "bias" will calibrate the force sensor by measuring the baseline voltage arising from no load on the force sensor. In the next step, the signal arising from different known weights is measured. Equation (1) gives the voltage-to-milligram transition coefficient C:

$$C = \frac{m_w}{U_w - U_B} \tag{1}$$

where, m_w is the weight of the calibration load in milligrams, U_w is the voltage of the force sensor's output which is measured by voltmeter when the calibration load is applied, and U_B is the bias voltage. The coefficient is stored in a .txt file. In order to calculate the force from raw voltage measurements, Equation (2) is presented:

$$F = C(U - U_B) \tag{2}$$

where, F is the calculated force, and U is the potential measured from the force sensor's output.

2.4.2. Elasticity Measurements

Elasticity determination is essential to obtain linear length change (strain) results. In all available commercial muscle analyzers, only the calibration of the force sensor is performed by a defined weight, which will provide uncertainties in real measurements due to the fact that the elastic modulus of an ionic actuator changes during charging/discharging cycles [34,35]. The elasticity estimation of an artificial muscle is based on Hooke's law (Equation (3)):

$$F = -kX \tag{3}$$

where, F is the force needed to extend or compress an object by some distance X and k is a constant factor characterizing object elasticity. If a derivative of the force is taken with respect to distance, Equation (4) is obtained:

$$k = -\frac{\partial F}{\partial X} \tag{4}$$

The idea was to perform a known movement with LAS and measure the changes of force. Figure S2 shows the block diagram of the software with the respective steps in the description of the algorithm.

2.4.3. Initialization and Experiments with the Potentiostat

The potentiostat controls the electrochemical processes of ionic actuators, which translate the electrical signal (potential, current) to shape change, which is measured as force (weight change) or strain (length change).

The initialization process of the potentiostat is based on three steps: (a) reading the variables describing how the experiment will be performed, (b) setting up the potentiostat for the experiment, and (c) starting the experiment. The setup variables are directly read from the user interface from user-controllable variables. The experiment type is selected by the user from a set of three: cyclic voltammetry (CV), chronoamperometry (CA), and chronopotentiometry (CP). The block diagram for the potentiostat is shown in Figure S3a, strain and stress measurements have the same basic structure.

2.4.4. Strain and Stress Measurements

Strain experiments also include an additional muscle length controlling logic, as the upper clamp is moved by the LAS plate to maintain constant force during the scan. The magnitude of the required movement by the LAS constitutes the isotonic strain. In order to keep a constant force, a proportional controller (P controller) is implemented, as presented in Equation (5), which derives from Equation (4):

$$\Delta X = -k' \Delta F \tag{5}$$

where, the gain $k' = (k + u_k)$, and u_k is the uncertainty of k. The addition of u_k to k assures that no faulty steps are done in a case when $k \cong u_k$. The input (ΔF) is the difference between the measured force and the set state (force value which the P controller is intended to track). The output is a step size (ΔX) needed to maintain the muscle in the set state of force. The gain (k) for the P controller is the elasticity coefficient of a muscle, whose strain is to be measured. The minimum steps size that the LAS can carry out is $\Delta l_{min} = 0.5$ µm. In case the ΔF is smaller then:

$$\Delta F_{min} = \frac{\Delta l_{min}}{-k'} \tag{6}$$

No movement is carried out since it would require a step size $< \Delta l_{min}$. In such cases, the electroactive material has linear displacement below the resolution of the LAS, the actual force can differ from the set value and is equal to ΔF_{min}. The detailed description of the block diagram for strain measurement is given in Figure S3b.

The automated multiple stress or strain measurement process is shown in Figure 2a and the user interface of the isotonic (displacement) and isometric (force) electro-chemo-measurement software (IIECMS) [33] is presented in Figure 2b. The IIECMS program is designed for measurement automation with the additional function of avoiding user errors in mislabeling or accidentally selecting wrong settings for the experiments.

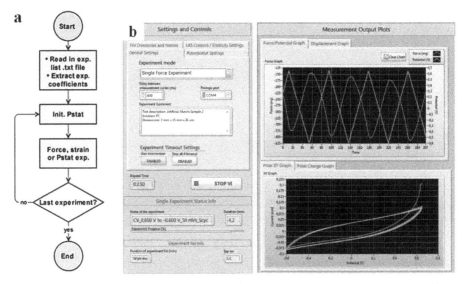

Figure 2. a: Block diagram of the automated multiple stress or strain measurement processes, and **b:** Layout of the graphical user interface (single force experiment) of the developed software.

The block diagram of the automated measurement process (Figure 2a) reads the experiment list and coefficients from a previously generated .txt file. Coefficients, which describe the experiments, are extracted and placed into a temporary array. The program enters into a for-loop, where the potentiostat is initialized and the user-selected measurement is executed. The number of cycles in the for-loop is equal to the amount of experiments on the list.

The graphical user interface (GUI) in Figure 2b shows three main sections. Section 1 includes all controls, such as general settings, file direction and names, potentiostat settings, and motor controls/elasticity measurement. Section 2 includes a set of indicators of experiment progress, such as ongoing experiment and experiment list info's. In Section 3, four graphs are included that plot measurement data.

3. Results on Model Systems and Discussion

The IIECMS program was adapted on real sample measurements. Two different types of test samples were measured (conducting polymers PPy/DBS and MWCNT-CDC fibers) to show isotonic and isometric measurement results. In the case of the conducting polymer, the isotonic length change or displacement amplitude was more than 20 times higher (referred to as high signal-to-noise ratio ca. 20 dB) than the actuation resolution of the LAS ($\Delta l_{min} = 0.5$ µm). The second measurement represents a situation where the displacement amplitude is comparable to Δl_{min} (referred to as low signal-to-noise ratio) and a measurement resolution refining technique is described.

Over 50 different isometric and isotonic measurements were made on ionic material samples, meaning about 200 h of measurement time. The minimum data acquisition period of this measurement setup was 160 ms.

3.1. Characterization of Ionic Actuator Materials

In the case of PPy/DBS, the DBS⁻ counterions are immobilized during electropolymerization [36] and upon discharging, their negative charge is compensated by (solvated) cations, as shown before [37,38]. In the case of MWCNT-CDC fibers, the actuation mechanism is based on the charging of the electric double layer (non-Faradaic actuator) [10]. Figure 3 shows the scanning electron microscopy (SEM) images of the two different electroactive materials applied in this work.

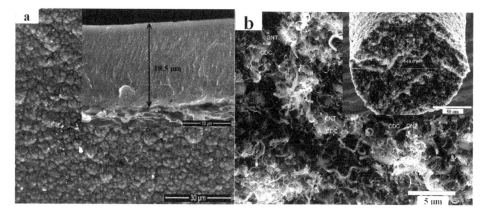

Figure 3. Scanning electronic microscopy (SEM) images of **a**: Polypyrrole Doped with Dodecylbenzene-Sulfonate (PPy/DBS) surface (scale bar 30 µm) with inset—the cross-section (scale bar 10 µm), and **b**: MWCNT-CDC fiber surface (scale bar 5 µm) with inset—cross-section (scale bar 50 µm).

The surface of the PPy/DBS film (Figure 3a) showed a typical cauliflower structure [39], with the cross-section showing a thickness of 18.5 µm. The diameter of the MWCNT-CDC fiber, as measured from the cross section (inset in Figure 3b), was 149.6 µm, the solid particles partly seen represent CDC surrounded by MWCNT material [10] (marked in Figure 3b). The conductivity of the PPy/DBS films was $0.5 \pm 0.04\,\text{S cm}^{-1}$ while MWCNT-CDC fiber had $13.5 \pm 7\,\text{S cm}^{-1}$, in line with those shown before [10]. When it comes to linear actuation measurements, the mechanical properties, such as brittleness and elasticity, are important [13]. PPy/DBS films were easy to handle for fixing them on the force sensor and the upper clamp (Figure 2a), while the MWCNT-CDC fiber was very brittle and required extremely delicate handling.

3.2. High Signal-to-Noise Ratio Isotonic and Isometric Measurements

The performance of the setup was demonstrated on PPy/DBS samples in LiTFSI-PC solution in the potential range of 0.65 to −0.6 V under isometric (Figure 4a) and isotonic (Figure 4b) modes. This relates to case one, where isotonic length change or displacement amplitude is about 2 orders of magnitude larger than the minimal step size of the LAS ($\Delta l_{min} = 0.5\mu m$).

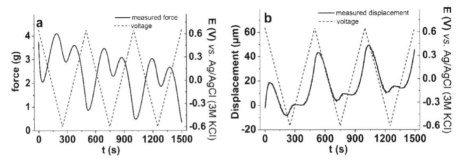

Figure 4. Cyclic voltammetry measurements (scan rate $5\,\text{mV s}^{-1}$, 3 cycles) of PPy/DBS films in LiTFSI-PC electrolyte in voltage potential (dashed) range 0.65 to -0.6V showing in **a**: isometric (measured force, black line), and **b**: isotonic (measured displacement, black line) results against time.

Figure 4a shows the force (weight change) results of the isometric measurements of PPy/DBS films driven by cyclic voltammetry. The maximum stress for this film was found in range of 0.9 MPa. In the case of PPy/DBS, which in general is a typical cation-driven actuator due to the immobile DBS⁻ ions left in the PPy network, the actuation in LiTFSI-PC is a special phenomenon where the DBS⁻Li⁺ ion pairs partly become undissociated in propylene carbonate solvent due to its aprotic nature [13,40]. Therefore, new places of the PPy/DBS film are oxidized and the solvated counter ions TFSI⁻ can enter the film and swell the film at oxidation (expansion, seen in Figure 4b). Our setup was easily able to detect and distinguish between the two processes taking place during one charging cycle. During the first scan (Figure 4a), the potential went from 0.65 to −0.6 V and the force decreased, then increased and finally showed a small decrease at −0.65 V. The contraction corresponds to a small cation involvement during reduction (seen in Figure 4b of displacement measurements, maximum strain in range of 0.6%). During the next cycles, the force was reduced, corresponding to increased displacement (strain). In both displacement and force graphs, creep [24] can be detected, which means that the "neutral" position of the linear PPy/DBS actuator changes, a relatively common behavior for ionic electroactive materials [41]. The phase shift (Figure 4b) of displacement in respect to the driving signal voltage is due to the participation of cations, which in the initial phase of a cycle lead to volume contraction, before the anions take over [13].

Upon very careful observation, one can see some noise appearing in Figure 4b, for example at position 200 s, 700 s, and 1200 s. The noise is introduced when the actual elasticity coefficient, which is used for P controller, temporarily differs from the measured value. This is due to the ion−matrix interactions. In extreme cases, when the elasticity coefficient is measured to be much lower, the P controller (Equation (6)) starts to generate. The elastic coefficient, k, was calculated from

Polymers **2019**, *11*, 1054

Equation (3), giving for PPy/DBS, before charging/discharging, the value of 239 mg/µm and after charging/discharging (50 actuation cycles), the value of 134 mg/µm. As it has been shown previously, the elastic modulus of conducting polymers changes during actuation cycles [42] was also effected by the nature of the solvent applied [43]. Therefore, as seen from the above results, the determination of the elastic coefficient to operate isotonic measurements is needed in order to measure meaningful data.

3.3. Low signal-to-Noise Ratio (SNR) Isotonic Measurements

To examine materials which have very low displacement amplitude, comparable to Δl_{min} (near 0 dB, SNR) MWCNT-CDC fibers were applied in isotonic displacement measurements conducted with cyclic voltammetry (scan rate 5 mV s^{-1}) but with the same electrolyte and potential range seen in Figure 4b. The results are presented in Figure 5.

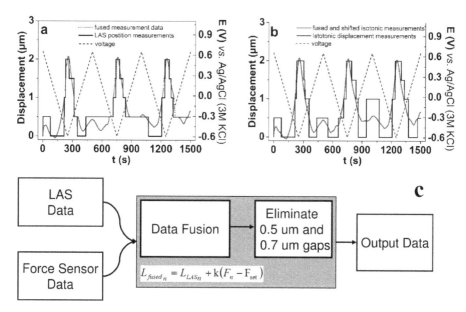

Figure 5. MWCNT-CDC fiber under cyclic voltammetry (scan rate 5 mV s^{-1} (3 cycles), potential range 0.65 to −0.6 V) in LiTFSI-PC electrolyte. **a**: Near 0 dB SNR isotonic displacement measurement results (black line) and force sensor data fused with measurements data (red points) with voltage (dashed, blue). **b**: Fused and shifted isotonic measurements (red points) after being processed by slack correction algorithm with voltage (dashed, blue) compared to isotonic displacement (original LAS position measurements). **c**: Near 0 dB SNR strain measurement data correction algorithm. LAS and force sensor data (inputs) are fused via Equation (7). Then 0.5 um and 0.7 um gaps are eliminated, and resulted data is outputted.

Figure 5 shows the case where the displacement (isotonic measurements) of the ionic electroactive material is in the limitation of the measurement setup. Figure 5a shows the raw measurements of MWCNT-CDC actuators under the cyclic voltammetric technique, which shows a rough estimate about the specimen's isotonic actuation properties. The graph in Figure 5a in the present form cannot be used in further data analysis. To give a better interpretation of the results, the force sensor data can be fused with isotonic displacement measurements. We know from the elastic coefficient measurements that the change in force is proportional to the change of length, therefore the fusion can be implemented as Equation (7):

$$L_{fused_n} = L_{LAS_n} + k(F_n - F_{set}) \tag{7}$$

where, n is the index of the measurement data sample, L_{LAS_n} is the isotonic displacement measured by LAS's controller, F_n is the measured force, and F_{set} is a set force which is constant throughout the experiment. Taking a closer look at the fused measurements shown in Figure 5a, there is some sort of ambiguous slack in the LAS actuation system, for example in the range of 310 s to 320 s, in the form of a 0.5 μm gap (comparable with Δl_{min}) between the measurement points, whereas similar gaps appear at other time steps as well. We assume that the slack might be a combination of the slack in the muscle samples clamping system, the internal properties of the samples, and the LAS controller position measurement or actuation errors. We assume that since the combined slack is comparable with Δl_{min}, it is most likely caused by the LAS controller error. Figure 5b shows the fused isotonic displacement (red points) after being processed by a slack correction algorithm in comparison to the original LAS position measurements. Figure 5c gives the data correction algorithm in the near 0 dB SNR displacement measurements by using two cycles, where the first removed the 0.5 μm gaps and the second the 0.7 μm gaps. After applying this algorithm, the results of the displacement measurements of MWCNT-CDC fibers can be interpreted.

Figure 5a,b results in expansion at discharging (−0.65 V) in the propylene carbonate solvent, which was explained [10] by the anions being nearly immobile, whereas at discharging the (solvated) cations (Li⁺) balance the negative charge bringing along the length change of the actuator [44]. In the case of the MWCNT-CDC fiber studied in this research, the main expansion also appeared at discharging with displacement in the range of 2 μm (equivalent to 0.2% strain). It was also found that small expansion at oxidation in the range of 0.3 μm appeared, which we assume relies on a small expansion accompanying the charging process of the EDL formed [45].

3.4. Uncertainty Evaluation

The uncertainty of force measurements depends on the components [46] which are used in the force calculations. Since force is calculated based on Equation (2), the combined uncertainty is shown in Equation (8):

$$\mu = \sqrt{\left(\frac{\partial F}{\partial C}u_c\right)^2 + \left(\frac{\partial F}{\partial U_B}u_{U_B}\right)^2 + \left(\frac{\partial F}{\partial U}u_U\right)^2 + u_{Fd}^2} \tag{8}$$

where, u_C is the uncertainty of the voltage-to-milligram coefficient, u_{U_B} is the uncertainty of the bias voltage, u_U is the uncertainty of the potential measured from the force sensor, and u_{Fd} is the uncertainty caused by the drift of force measurements. The expanded uncertainty of stress and strain measurements is stated as the standard uncertainty of measurement multiplied by the coverage factor $k = 2$, which for a normal distribution corresponds to a coverage probability of approximately 95%. The uncertainty of the force measurements is proportional to the voltage values measured from the force sensor, which can be seen from Equation (2). This means that small force amplitude signals have higher accuracy, whereas with the growth of signal amplitude, the measurement uncertainty increases. Figure 6 shows the case of small amplitudes in force (mg) for MWCNT-CDC fibers.

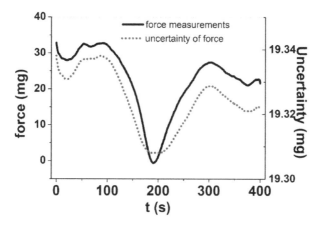

Figure 6. Cyclovoltammetric (scan rate 5 mV s^{-1}, 3rd cycle) isometric measurement results of MWCNT-CDC fibers in LiTFSI-PC electrolyte at potential range 0.65 to -0.6V, showing the force measurements (mg) (black line) and the uncertainty (B type) of the force measurements (mg) (red dotted).

The MWCNT-CDC fiber (Figure 6) showed change in force in the range of 30.5 mg, which is translated to stress in the range of 17 kPa. There is a small decrease in force from the starting point of 33 mg to the end point (22.8 mg) of 10.2 mg. The decrease in maximum force belongs to the creep effect, which can appear if mixed ion involvement appears, which is also seen in recent research [47].

The discharging, therefore, leads to the decrease in force which is translated into displacement knowing the elastic coefficient of the MWCNT-CDC fiber (k = 13.5 mg/μm). According to the uncertainty measurement in Figure 6 based on Equation (8), the coefficient u_c was estimated to be 16.8 g/V and $u_{UB} = u_U = 0.005$ mV. The most important contributor to displacement measurement error is the resolution of the LAS position estimation system, which is 0.5 μm, therefore, the B type uncertainty of strain measurements without the assistance of data fusing (Equation (8)) is $k\frac{0.5\ \mu m}{\sqrt{3}} \cong 0.6$ μm. The data fusion contributes marginally (usually <0.05 μm) to the uncertainty of strain measurements, but it clarifies the interpretation of the results.

4. Conclusions

A custom software solution was designed and implemented for driving and measuring ionic elelctroactive material-based systems. The hardware setup combined separately acquired commercial off-the-shelf devices like a force sensor, a voltmeter, a linear actuation stage, and a potentiostat. To operate the electro-chemo measurement system in a synchronized manner, a new software with graphical interface was developed in LabVIEW. The IIECMS was made by combining software development kits of the implemented devices, creating algorithms for the execution of isotonic and isometric measurements, and automating the measurement and user input processes in LabVIEW. The maximum user-selectable measurement frequency for IIECMS was 160 ms. In order to operate the potentiostat in the LabVIEW, a respective set of software drivers were created with ActiveX-based SDK of the potentiostat. The IIECMS program was applied on PPy/DBS linear actuators revealing the first case of 20 dB high signal-to-noise ratio isotonic and isometric measurements with maximum force of 3 g (stress of 0.9 MPa) and strain in the range of 4.3% at oxidation. The system was easily capable of detecting mixed-mode driving and distinguishing between the cation and the anion flux induced actuation. The MWCNT-CDC fibers belonged to the second case of 0 dB low signal-to-noise ratio isotonic measurements. Partial discontinuity of the data indicated that the IIECMS setup has an occasionally appearing slack of approximately 0.7 μm, which can be eliminated with post experiment data processing. The displacement of the MWCNT-CDC fibers was found in the limitation of Δl_{min}

(0.5 μm) with maximum displacement at a discharging of 2 μm (0.2 % strain). The accuracy of isotonic measurement was estimated to be in the range of 0.6 μm, with the largest contributor relying on the measurement error of the position estimation resolution of LAS. Fusing the output of different channels allows for successful operation even at such challenging conditions.

Future perspectives for the IIECMS program will be to fully automate the isotonic and isometric experiments. This means that the user has an option to run both experiment types in an automated sequence, thus eliminating the need for a user to measure sample elasticity. In principle, any type of voltage- or current-driven linear actuators in the form of strips, fibers, etc., could be studied with the IIECMS program.

Supplementary Materials: The following are available online at http://www.mdpi.com/2073-4360/11/6/1054/s1. Figure S1 General structure of the developed software, Figure S2 Block diagram of the elasticity measurement process, Figure S3 Block diagram of potentiostat and strain measurement.

Author Contributions: Conceptualization, T.T.; Investigation, M.H.; Software, G.A.; Supervision, R.K.

Acknowledgments: The work was supported by the Estonian Research Council Grant IUT20-24.

Conflicts of Interest: The authors declare no conflict of interest.

References

1. Kosidlo, U.; Omastová, M.; Micusík, M.; Ćirić-Marjanović, G.; Randriamahazaka, H.; Wallmersperger, T.; Aabloo, A.; Kolaric, I.; Bauernhansl, T. Nanocarbon based ionic actuators—A review. *Smart Mater. Struct.* **2013**, *22*, 104022. [CrossRef]
2. Mirfakhrai, T.; Madden, J.D.W.; Baughman, R.H. Polymer artificial muscles. *Mater. Today* **2007**, *10*, 30–38. [CrossRef]
3. Yan, B.; Wu, Y.; Guo, L. Recent advances on polypyrrole electroactuators. *Polymers* **2017**, *9*, 446. [CrossRef] [PubMed]
4. Daneshmand, M.; Abels, A.; Anbarjafari, G. Real-time, automatic digi-tailor mannequin robot adjustment based on human body classification through supervised learningNo Title. *Int. J. Adv. Robot. Syst.* **2017**, *13*, 1–9.
5. Maziz, A.; Concas, A.; Khaldi, A.; Stålhand, J.; Persson, N.-K.; Jager, E.W.H. Knitting and weaving artificial muscles. *Sci. Adv.* **2017**, *3*, 1–12. [CrossRef] [PubMed]
6. Torop, J.; Aabloo, A.; Jager, E.W.H. Novel actuators based on polypyrrole/carbide-derived carbon hybrid materials. *Carbon N. Y.* **2014**, *80*, 387–395. [CrossRef]
7. Le, T.H.; Kim, Y.; Yoon, H. Electrical and electrochemical properties of conducting polymers. *Polymers* **2017**, *9*, 150. [CrossRef]
8. Tomczykowa, M.; Plonska-Brzezinska, M. Conducting Polymers, Hydrogels and Their Composites: Preparation, Properties and Bioapplications. *Polymers* **2019**, *11*, 350. [CrossRef]
9. Cortés, M.T.; Moreno, J.C. Artificial muscles based on conducting polymers. *e-Polymers* **2003**, *3*, 1–42.
10. Plaado, M.; Kaasik, F.; Valner, R.; Lust, E.; Saar, R.; Saal, K.; Peikolainen, A.; Aabloo, A.; Kiefer, R. Electrochemical actuation of multiwall carbon nanotube fiber with embedded carbide-derived carbon particles. *Carbon N. Y.* **2015**, *94*, 911–918. [CrossRef]
11. Martinez, J.G.; Otero, T.F.; Jager, E.W.H. Effect of the electrolyte concentration and substrate on conducting polymer actuators. *Langmuir* **2014**, *30*, 3894–3904. [CrossRef] [PubMed]
12. Kiefer, R.; Martinez, J.G.; Kesküla, A.; Anbarjafari, G.; Aabloo, A.; Otero, T.F. Chemical Polymeric actuators: Solvents tune reaction-driven cation to reaction-driven anion actuation. *Sens. Actuar. B. Chem.* **2016**, *233*, 328–336. [CrossRef]
13. Kivilo, A.; Zondaka, Z.; Kesküla, A.; Rasti, P.; Tamm, T.; Kiefer, R. Electro-chemo-mechanical deformation properties of polypyrrole/dodecylbenzenesulfate linear actuators in aqueous and organic electrolyte. *RSC Adv.* **2016**, *6*, 69–75. [CrossRef]
14. Zhang, H.; Tang, J.; Zhu, P.; Ma, J.; Qin, L.C. High tensile modulus of carbon nanotube nano-fibers produced by dielectrophoresis. *Chem. Phys. Lett.* **2009**, *478*, 230–233. [CrossRef]
15. Bard, A.J.; Faulkner, L.R. Fundamentals and applications. *Electrochem. Methods* **2001**, *2*, 482.

16. Baughman, R.H.; Cui, C.; Zakhidov, A.A.; Iqbal, Z.; Barisci, J.N.; Spinks, G.M.; Wallace, G.G.; Mazzoldi, A.; De Rossi, D.; Rinzler, A.G.; et al. Carbon nanotube actuators. *Science* **1999**, *284*, 1340–1344. [CrossRef] [PubMed]

17. Hantel, M.M.; Presser, V.; Kötz, R.; Gogotsi, Y. In situ electrochemical dilatometry of carbide-derived carbons. *Electrochem. Commun.* **2011**, *13*, 1221–1224. [CrossRef]

18. Torop, J.; Arulepp, M.; Leis, J.; Punning, A.; Johanson, U.; Palmre, V.; Aabloo, A. Nanoporous Carbide-Derived Carbon Material-Based Linear Actuators. *Materials* **2010**, *3*, 9–25. [CrossRef]

19. Carpi, F.; Chiarelli, P.; Mazzoldi, A.; De Rossi, D. Electromechanical characterisation of dielectric elastomer planar actuators: Comparative evaluation of different electrode materials and different counterloads. *Sens. Actuar. A Phys.* **2003**, *107*, 85–95. [CrossRef]

20. Spinks, G.M.; Xi, B.; Zhou, D.; Truong, V.T.; Wallace, G.G. Enhanced control and stability of polypyrrole electromechanical actuators. *Synth. Met.* **2004**, *140*, 273–280. [CrossRef]

21. Spinks, G.M.; Liu, L.; Wallace, G.G.; Zhou, D. Strain response from polypyrrole actuators under load. *Adv. Funct. Mater.* **2002**, *12*, 437–440. [CrossRef]

22. Kaneto, K.; Fujisue, H.; Kunifusa, M.; Takashima, W. Conducting polymer soft actuators based on polypyrrole films—Energy conversion efficiency. *Smart Mater. Struct.* **2007**, *16*, S250–S255. [CrossRef]

23. Madden, J.D.; Cush, R.A.; Kanigan, T.S.; Brenan, C.J.; Hunter, I.W. Encapsulated polypyrrole actuators. *Synth. Met.* **1999**, *105*, 61–64. [CrossRef]

24. Madden, J.D.; Rinderknecht, D.; Anquetil, P.A.; Hunter, I.W. Creep and cycle life in polypyrrole actuators. *Sens. Actuar. A Phys.* **2007**, *133*, 210–217. [CrossRef]

25. Xiao, T.; Ren, Y.; Liao, K.; Wu, P.; Li, F.; Cheng, H.M. Determination of tensile strength distribution of nanotubes from testing of nanotube bundles. *Compos. Sci. Technol.* **2008**, *68*, 2937–2942. [CrossRef]

26. ADInstruments. Available online: https://www.adinstruments.com/products/labchart (accessed on 16 June 2019).

27. Scientific, A. Available online: https://aurorascientific.com/products/muscle-physiology/muscle-software/dynamic-muscle-control/ (accessed on 16 June 2019).

28. Panlab. Available online: https://www.panlab.com/en/products/protowin-software-panlab (accessed on 16 June 2019).

29. Della Santa, A.; De Rossi, D.; Mazzoldi, A. Performance and work capacity of a polypyrrole conducting polymer linear actuator. *Synth. Met.* **1997**, *90*, 93–100. [CrossRef]

30. Instruments, N. Available online: http://www.ni.com/labview/ (accessed on 16 June 2019).

31. Scientific, A. Available online: https://aurorascientific.com/products/muscle-physi (accessed on 16 June 2019).

32. Software, Acquisition, ACAD®Isolated Tissue Data. Available online: https://www.harvardapparatus.co.uk/webapp/wcs/stores/servlet/product_11555_10001_52805_-1_HAUK_ProductDetail__ (accessed on 16 June 2019).

33. Valner, R. Software Development for Isotonic and Isometric Electro- Chemo Measurement System. 2015. Available online: https://pdfs.semanticscholar.org/551f/09a5410462ac9bb339ead3864c29af94468b.pdf (accessed on 16 June 2019).

34. Christophersen, M.; Shapiro, B.; Smela, E. Characterization and modeling of PPy bilayer microactuators. *Sens. Actuar. B Chem.* **2006**, *115*, 596–609. [CrossRef]

35. Khadka, R.; Aydemir, N.; Kesküla, A.; Tamm, T.; Travas-Sejdic, J.; Kiefer, R. Enhancement of polypyrrole linear actuation with poly(ethylene oxide). *Synth. Met.* **2017**, *232*, 1–7. [CrossRef]

36. Wang, X.; Smela, E. Experimental Studies of Ion Transport in PPy(DBS). *J. Phys. Chem. C* **2009**, *113*, 369–381. [CrossRef]

37. Aydemir, N.; Kilmartin, P.A.; Travas-Sejdic, J.; Kesküla, A.; Peikolainen, A.-L.; Parcell, J.; Harjo, M.; Aabloo, A.; Kiefer, R. Electrolyte and solvent effects in PPy/DBS linear actuators. *Sens. Actuar. B Chem.* **2015**, *216*, 24–32. [CrossRef]

38. Vidanapathirana, K.P.; Careem, M.A.; Skaarup, S.; West, K. Ion movement in polypyrrole/dodecylbenzenesulphonate films in aqueous and non-aqueous electrolytes. *Solid State Ionics* **2002**, *154*, 331–335. [CrossRef]

39. Gade, V.K.; Shirale, D.J.; Gaikwad, P.D.; Kakde, K.P.; Savale, P.A.; Kharat, H.J.; Shirsat, M.D. Synthesis and Characterization of Ppy-PVS, Ppy-pTS, and Ppy-DBS Composite Films. *Int. J. Polym. Mater.* **2007**, *56*, 107–114. [CrossRef]

40. Kiefer, R.; Kesküla, A.; Martinez, J.G.; Anbarjafari, G.; Torop, J.; Otero, T.F. Interpenetrated triple polymeric layer as electrochemomechanical actuator: Solvent influence and diffusion coefficient of counterions. *Electrochim. Acta* **2017**, *230*, 461–469. [CrossRef]

41. Tominaga, K.; Hamai, K.; Gupta, B.; Kudoh, Y.; Takashima, W.; Prakash, R.; Kaneto, K. Suppression of electrochemical creep by cross-link in polypyrrole soft actuators. *Phys. Procedia* **2011**, *14*, 143–146. [CrossRef]

42. Smela, E.; Gadegaard, N. Volume change in polypyrrole studied by atomic force microscopy. *J. Phys. Chem. B* **2001**, *105*, 9395–9405. [CrossRef]

43. Khadka, R.; Zondaka, Z.; Kesküla, A.; Safaei Khorram, M.; Thien Khanh, T.; Tamm, T.; Travas-Sejdic, J.; Kiefer, R.; Minh City, C. Influence of solvent on linear polypyrrole-polyethylene oxide actuators. *J. Appl. Polym. Sci.* **2018**, *46831*, 1–7. [CrossRef]

44. Hahn, M.; Barbieri, O.; Campana, F.P.; Kötz, R.; Gallay, R. Carbon based double layer capacitors with aprotic electrolyte solutions: The possible role of intercalation/insertion processes. *Appl. Phys. A Mater. Sci. Process.* **2006**, *82*, 633–638. [CrossRef]

45. Kaasik, F.; Tamm, T.; Hantel, M.M.M.; Perre, E.; Aabloo, A.; Lust, E.; Bazant, M.Z.Z.; Presser, V. Anisometric charge dependent swelling of porous carbon in an ionic liquid. *Electrochem. Commun.* **2013**, *34*, 196–199. [CrossRef]

46. Xu, C. A practical model for uncertainty evaluation in force measurements. *Meas. Sci. Technol.* **1998**, *9*, 1831–1836. [CrossRef]

47. Kiefer, R.; Aydemir, N.; Torop, J.; Tamm, T.; Temmer, R.; Travas-Sejdic, J.; Must, I.; Kaasik, F.; Aabloo, A. Carbide-derived carbon as active interlayer of polypyrrole tri-layer linear actuator. *Sens. Actuar. B Chem.* **2014**, *201*, 100–106. [CrossRef]

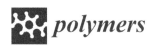

Article

Enhanced Stability and Driving Performance of GO–Ag-NW-based Ionic Electroactive Polymer Actuators with Triton X-100-PEDOT:PSS Nanofibrils

Minjeong Park [1], Seokju Yoo [1], Yunkyeong Bae [1], Seonpil Kim [2] and Minhyon Jeon [1,*]

[1] Department of Nanoscience and Engineering, Center for Nano Manufacturing, Inje University, Gimhae 50834, Korea; mjpark9121@gmail.com (M.P.); yuinjae7@gmail.com (S.Y.); yunkyeong.b@gmail.com (Y.B.)

[2] Department of Military Information Science, Gyeongju university, Gyeongju 38065, Korea; seonpil@gu.ac.kr

* Correspondence: mjeon@inje.ac.kr; Tel.: +82-55-320-3672; Fax: +82-55-320-3963

Received: 25 April 2019; Accepted: 16 May 2019; Published: 19 May 2019

Abstract: Ionic electroactive polymers (IEAPs) have received considerable attention for their flexibility, lightweight composition, large displacement, and low-voltage activation. Recently, many metal–nonmetal composite electrodes have been actively studied. Specifically, graphene oxide–silver nanowire (GO–Ag NW) composite electrodes offer advantages among IEAPs with metal–nonmetal composite electrodes. However, GO–Ag NW composite electrodes still show a decrease in displacement owing to low stability and durability during driving. Therefore, the durability and stability of the IEAPs with metal–nonmetal composite electrodes must be improved. One way to improve the device durability is coating the electrode surface with a protective layer. This layer must have enough flexibility and suitable electrical properties such that it does not hinder the IEAPs' driving. Herein, a poly(3,4-ethylenedioxythiophene)–poly(styrenesulfonate) (PEDOT:PSS) protective layer and 4-(1,1,3,3-tetramethylbutyl)phenyl-polyethylene glycol (Triton X-100) are applied to improve driving performance. Triton X-100 is a nonionic surfactant that transforms the PEDOT:PSS capsule into a nanofibril structure. In this study, a mixed Triton X-100/PEDOT:PSS protective layer at an optimum weight ratio was coated onto the GO–Ag NW composite-electrode-based IEAPs under various conditions. The IEAP actuators based on GO–Ag NW composite electrodes with a protective layer of PEDOT:PSS treated with Triton X-100 showed the best stability and durability.

Keywords: graphene oxide; silver nanowires; ionic electroactive polymer; poly(3,4-ethylenedioxythiophene)–poly(styrenesulfonate) (PEDOT:PSS); 4-(1,1,3,3-Tetramethylbutyl)phenyl-polyethylene glycol

1. Introduction

Ionic electroactive polymers (IEAPs) are among the most functional materials-based actuators. IEAPs have useful properties, such as a lightweight composition, a large working displacement under a low driving voltage, and a high energy density [1]. Electrodes are an important component of IEAPs. When sufficient, they provide high electrical conductivity, mechanical durability, and a smooth surface morphology [2]. The electrodes of IEAPs are categorized as metallic or nonmetallic. Noble metals (e.g., platinum, gold) with high electrical conductivity and electrochemical stability are often used as the metallic electrode of IEAPs. However, most metallic electrodes exhibit microcracks on the electrode surface, thus diminishing their surface electrical conductivity during long-term actuation [3,4]. Meanwhile, most nonmetallic electrodes are fabricated using transition metal oxides or carbon materials [1,5–9]. These materials can be assembled into electrodes via a physical hot-pressing method, which is simpler and faster than other fabrication methods [10,11]. However, nonmetallic electrodes are less conductive than metallic electrodes. To resolve such challenges, recent studies have increasingly investigated metallic–nonmetallic composite electrodes. Composite electrodes can solve both the occurrence of microcracks on the metallic electrode surface during driving and the low electrical conductivity of nonmetallic electrodes [12]. However, composite electrodes have a high contact resistance on their surfaces, which can instigate electrode burn out and decrease electrode durability during actuation.

Poly(3,4-ethylenedioxythiophene)–poly(styrenesulfonate) (PEDOT:PSS), a conductive polymer, is applied in various fields for its advantages of electrical conductivity and transparency. Particularly, PEDOT:PSS has been actively studied as a protective layer coated onto electrode materials such as carbon nanotubes (CNTs), graphene, and metal nanowires (NWs) with high contact resistance [13–16]. PEDOT:PSS is composed of a PEDOT phase and a PSS phase. The PEDOT phase is electrically conductive, but it is not suitable for solution processing because of its low solubility. On the other hand, the PSS phase is electrically insulating and water soluble. Because of these combined properties, the PEDOT:PSS is both electrically conductive and water soluble. However, the PEDOT phase's tendency to aggregate can make electrodes brittle.

Triton X-100 (4-(1,1,3,3-tetramethylbutyl)phenyl-polyethylene glycol) is composed of hydrophilic polyethylene oxide and hydrophobic 4-(1,1,3,3-tetramethylbutyl)-phenyl. This composition imparts it with amphiphilic properties owing to its hydrophilic "head" and hydrophobic "tail" and is thus a nonionic surfactant. It is widely used for mixing polar and nonpolar materials or lysing cells to extract proteins or organelles in biological fields through its surfactant property [17]. Moreover, Triton X-100 can lead to the formation of PEDOT nanofibrils in a viscoelastic medium due to its amphiphilic molecular structure of Triton X-100. Additionally, it can solve the PEDOT phase aggregation problem. Through this mechanism, the structure of PEDOT:PSS can be modified to have a flexible morphology. In addition, it enables improving the electrical conductivity by removing the PSS phase, which acts an insulator, through post-treatment [18–20].

In this study, Triton X-100-PEDOT:PSS-coated graphene oxide and silver NWs (TP/GO–Ag NWs) were fabricated as composite-electrode-based IEAPs. Herein, it is demonstrated that Triton X-100 can transform the aggregated structure of PEDOT:PSS into a nanofibril structure and thus improve electrical properties. Furthermore, the Triton X-100/PEDOT (TP) mixture coated on GO–Ag NW composite electrodes is used as a protective layer for the GO–Ag NW electrode. The improved characteristics of TP/GO–Ag-NW-based IEAPs were observed with a focus on electrical conductivity and driving properties.

2. Materials and Methods

2.1. Materials

PEDOT:PSS, Triton X-100, GO, and Ag NWs were used to fabricate the composite electrodes of the IEAPs. PEDOT:PSS (1.3 wt %) and Triton X-100 were purchased from Sigma-Aldrich, St. Louis, MO, US. GO was synthesized using Hummer's method [21]. The Ag NW solution was purchased from Duksan Hi-Metal, Ulsan, Korea. Nafion 117 (N117) and 20 wt % Nafion resin (Nafion solution) were purchased from the DuPont Company, Midland, MI, US. 1-Ethyl-3-methylimidazolium trifluoromethylsulfonate (EMIM-Otf), an ionic liquid (IL), was purchased from Merck KGaA, Darmstadt, Germany. A polyvinylidene difluoride (PVDF) membrane filter with a pore size of 0.20 μm and a diameter ∅ 47 mm was purchased from Hyundai Micro., Ltd, Seongnam, Korea.

2.2. Fabrication of IEAP Actuators Based on TP/GO–Ag NW Electrode

GO powder (330 mg) and deionized water (5 mL) were stirred at 250 rpm for 5 min in a beaker to prepare the GO solution. The GO and Ag NW solutions were mixed in a 1:2.5 volume ratio. This GO–Ag NW mixture was used to fabricate a composite paper electrode using a vacuum filtration system. This paper electrode was dried at 100 °C for 5 min in a vacuum oven. We performed ion exchange on the purchased Nafion following a previously-reported method [19]. The Nafion resin was painted directly onto both surfaces of the N117 membrane as an additive solution to paste the GO–Ag NW composite electrode to the Nafion membrane, forming an electrode/membrane/electrode structure. This structure was then hot-pressed at 0.1 MPa and 100 °C for 2 min. Thus, we obtained IEAPs based on GO–Ag NW electrodes. In addition, Triton X-100 was mixed with PEDOT:PSS at various volume ratios and spin-coated onto both surfaces of the IEAP actuator based on the GO–Ag NW composite electrodes. Two coating conditions were considered: duration and speed. The PSS phases were then removed with methanol. Finally, IEAPs based on TP/GO–Ag NW composite electrodes were fabricated (width × length (Free length) × thickness = 5 mm × 35 mm (30 mm) × 200 μm).

2.3. Characterization

The transmittances of samples with different Triton X-100 weight ratios were investigated using an ultraviolet visible spectrophotometer (LAMBDA 465, PerkinElmer, Seoul, Korea). The morphologies and thickness of the electrodes were investigated using atomic force microscope (AFM; XE-100, Park Systems, Suwon, Korea) and field-emission scanning electron microscopy (FE-SEM; S-4300, Hitachi, Tokyo, Japan). The sheet resistances of electrodes were measured using a four-point probe (FPP-HS 8, DASOL ENG, Cheongju, Korea). The driving characteristics of the actuators were measured using a laser displacement sensor (ZS-LD80, OMRON Korea, Seoul, Korea) in an actuation performance analyzer that we constructed.

3. Results

3.1. Characterization of the TP/GO–Ag NW Electrode

Triton X-100 and PEDOT:PSS were mixed at six different weight ratios (0.0, 1.0, 2.5, 5.0, 7.5, and 10.0 wt % Triton X-100). In addition, the PSS phases of all Triton X-100 and PEDOT:PSS mixtures were removed with methanol. In order to optimize the Triton X-100 concentration in the TP mixtures, protective layers were separately spin-coated on glass. Figure 1a shows the TP mixtures with different Triton X-100 contents coated on glass. Pure PEDOT:PSS did not uniformly coat the glass because of the high surface tension of PEDOT:PSS and its aggregated structure. However, the added Triton X-100 reduced the surface tension of the various TP mixtures. Thus, the TP mixtures uniformly coated the glass, in contrast with the pure PEDOT:PSS solution.

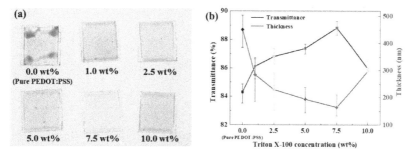

Figure 1. (**a**) The optical images of Triton X-100/poly(3,4-ethylenedioxythiophene) (TP) mixture coated on glass and (**b**) transmittance (black line) and thickness (blue line) of films with different Triton X-100 weight ratios.

Figure 1b shows the optical transmittance at 550 nm and the thickness of the pure PEDOT:PSS and TP mixtures coated on the glass. The transmittance and thickness of all samples are inversely proportional. Pure PEDOT:PSS had lower transmittance (84.29%) than the TP mixtures. Meanwhile, the TP mixture with 7.5 wt % Triton X-100 had the highest transmittance (88.81%) and the thickest coating (164.29 nm) among the TP mixtures. These results suggests that the aggregated structure of PEDOT:PSS transformed into a nanofibril structure through the addition of Triton X-100.

Figure 2a–f shows AFM images of the 0.0, 1.0, 2.5, 5.0, 7.5, and 10.0 wt % TP layers, respectively, which reveal the surface roughness and structure of the TP layer. For accurate comparison, we scanned and analyzed an area of 5×5 µm². These results reveal that the Triton X-100 treatment transformed the PEDOT:PSS aggregated structure to a nanofibril structure. Notably, the TP mixture with 7.5 wt % Triton X-100 had a nanofibril structure and lower surface roughness than the other TP mixtures, thus corroborating the transmittance and thickness results. At 10.0 wt %, the of the TP mixture exhibited aggregation, meaning that the amount of Triton X100 exceeded the critical micelle concentration (CMC), thereby forming micelles. The CMC, the surfactant concentration above which micelles form, is an important characteristic of a surfactant. Specifically, after reaching the CMC, any additional surfactants added to the system form micelles, which are large molecules formed by clusters of surfactant particles such as Triton X100. The size of the particles increases with increasing molecular aggregation. Figure 2g schematically illustrates the mechanism underlying the PEDOT:PSS transformation with the Triton X100 treatment.

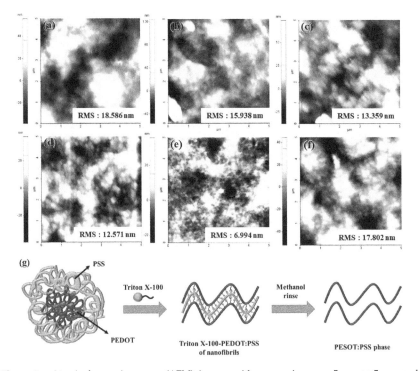

Figure 2. Atomic force microscope (AFM) images with a scanning area 5 µm × 5 µm and root-mean-square (RMS) surface roughness values. (**a**) 0.0 wt %, (**b**) 1.0 wt %, (**c**) 2.5 wt %, (**d**) 5.0 wt %, (**e**) 7.5 wt %, (**f**) 10.0 wt % Triton X100. (**g**) Schematic diagram illustrating the phase transition of poly(3,4-ethylenedioxythiophene)–poly(styrenesulfonate) (PEDOT:PSS) in the presence of Triton X-100.

Figure 3 graphically represents the sheet resistance of the GO–Ag NW composite electrodes coated with different TP mixtures, which was measured using a four-point probe. The TP-mixture-coated GO–Ag NW electrodes had a lower sheet resistance than the GO–Ag NW electrode without Triton X-100. In addition, washing with methanol decreased the sheet resistances of all the TP-mixture-coated GO–Ag NW electrodes. Particularly, the 7.5 wt % Triton X-100 TP mixture had the lowest sheet resistance. Table 1 lists the sheet resistance of the samples in detail. Accordingly, the 7.5 wt % Triton X-100 TP mixture is the most suitable choice as the protective layer for IEAP electrodes.

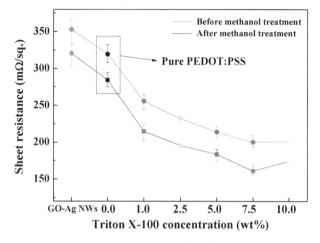

Figure 3. Sheet resistances before (red line) and after (black line) methanol treatment with various Triton X-100 weight ratios.

Table 1. Sheet resistances before and after methanol treatment for various Triton X-100 weight ratios.

Samples	Sheet Resistance (mΩ/sq.)	
	Before Methanol Treatment	After Methanol Treatment
GO–Ag NWs	352.65	320.25
0.0 wt % (Pure PEDOT:PSS)	319.36	284.16
1.0 wt % Triton X-100	255.52	214.73
2.5 wt % Triton X-100	232.15	195.57
5.0 wt % Triton X-100	213.80	183.57
7.5 wt % Triton X-100	200.08	161.00
10.0 wt % Triton X-100	201.00	174.00

Next, to optimize the spin coating conditions of the TP mixture with 7.5 wt % for the GO–Ag NW electrode, the TP/GO–Ag NW electrodes were fabricated with different coating times (0, 15, 30, and 45 s) and coating speeds (300, 500, 700, 1000, and 2000 rpm). Figure 4a shows the sheet resistance of the TP/GO–Ag NW electrodes with different coating times and a fixed coating speed of 1000 rpm. In this figure, the results for the TP-mixture-coated electrodes are compared with those of the GO–Ag NW electrode and the GO–Ag NW electrode coated with the pure PEDOT:PSS (P/GO–Ag NW electrode). The TP mixture with 7.5 wt % Triton X-100 coated for 30 s had a lower sheet resistance than the other samples. Figure 4b shows the sheet resistance of the TP/GO–Ag NW electrode at different coating speeds (coating time: 30 s). The 7.5 wt % Triton X-100 TP mixture coated at 1000 rpm for 30 s provided the lowest sheet resistance of about 161 mΩ/sq. Thus, the optimal spin coating conditions for coating the TP mixture on the GO–Ag NW electrode were 7.5 wt % Triton X100, a coating time of 30 s, and a coating speed of 1000 rpm. The sheet resistance of this electrode was 49.73% and 82.10% lower than those of the GO–Ag NW electrode and the P/GO–Ag NW electrode, respectively. These optimized conditions were thus used for subsequent experiments.

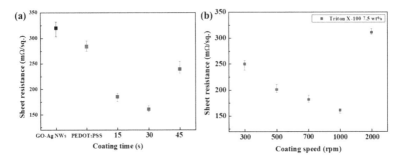

Figure 4. Sheet resistances of the (**a**) uncoated electrode (black point), P/GO–Ag NW electrode, and 7.5 wt % Triton X100 TP/GO–Ag NW electrodes coated for different coating times at 1000 rpm. (**b**) Sheet resistances of the 7.5 wt % Triton X100 TP/GO–Ag NW electrodes coated at different coating speeds.

3.2. Actuation Performance of IEAP Actuators Based on TP/GO–Ag NW Electrode

The driving performance of IEAP actuators with different electrodes was measured and observed, and the results are presented in Figure 5. The PEDOT:PSS-based IEAPs were measured to confirm the effect of 7.5 wt % Triton X-100. In order to investigate the effects of the electrode type on each actuator, the actuation performances of the IEAPs with three types of electrodes (GO–Ag NWs, P/GO–Ag NWs, and TP/GO–Ag NWs) were measured under ± 2.5 V_{AC} and 0.2 Hz, as shown in Figure 5a,b. Figure 5a shows the harmonic responses of the three types of IEAPs. The TP/GO–Ag-NW-based IEAPs had larger tip displacements than the other IEAPs.

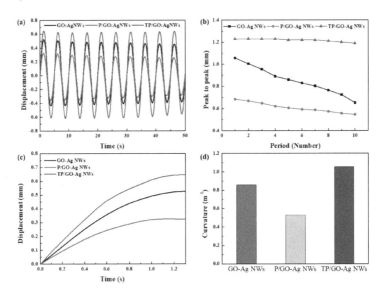

Figure 5. (**a**) Displacement versus time (± 2.5 V_{AC}, 0.2 Hz), (**b**) peak-to-peak performance, (**c**) response rate, and (**d**) bending curvature of three different ionic electroactive polymers (IEAPs) (based on GO–Ag NWs, P/GO–Ag NWs, and TP/GO–Ag NWs).

As shown in Figure 5b, the peak-to-peak performance of the TP/GO–Ag-NW-based IEAPs showed a lower slope than that of the other IEAPs, meaning that the TP/GO–Ag-NW-based IEAPs are more durable than the other IEAPs. Figure 5c shows the actuation performance, from 0 s to 1.3 s, of a segment of the harmonic response from Figure 5a. The response rate of the TP/GO–Ag-NW-based IEAPs was

34.83% and 23.87% faster than those with the P/GO–Ag-NW-based IEAPs and GO–Ag-NW-based IEAPs, respectively. Figure 5d shows the curvatures of the three types IEAPs. The maximum curvature of TP/GO–Ag-NW-based IEAPs was approximately 1.054 m^{-1}, which is higher than that of the GO–Ag-NW-based IEAPs (0.858 m^{-1}) and P/GO–Ag-NW-based IEAPs (0.53 m^{-1}).

After measuring the actuation performance, SEM was used to observe the change in the electrode surfaces of the three types IEAPs. The decreased actuation performance of the GO–Ag-NW- and P/GO–Ag-NW-based IEAPs in Figure 5a,b can be explained by Figure 6, which shows SEM images of the electrode surface of three types of IEAPs before (Figure 6a–c) and after (Figure 6d–f) the driving test. Because of the high contact resistance of the surface, heat is generated, which weakens metal NWs. This is a critical drawback of metal NWs networks, which may disconnect when voltage is administered to IEAPs owing to the resulting heat. Accordingly, serious transformation and network disconnection was observed for the Ag NWs in the IEAPs based on GO–Ag NWs and P/GO–Ag NWs. In contrast, the shape and network connection of the Ag NWs in the TP/GO–Ag-NW-based IEAPs were well maintained. Ultimately, this TP layer, which shows enhanced stability and durability during driving, can be used as a protective layer to decrease the high contact resistance of the GO–Ag NW electrode.

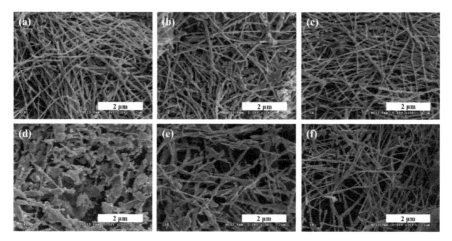

Figure 6. Surface scanning electron microscopy (SEM) images of the electrodes before (top) and after (bottom) a driving test of IEAPs based on (**a,d**) GO–Ag NWs, (**b,e**) P/GO–Ag NWs, and (**c,f**) TP/GO–Ag NWs.

4. Discussion and Conclusions

Triton X-100, a nonionic surfactant, was used to functionally enhance a PEDOT:PSS protective layer on a GO–Ag NW electrode. Triton X-100 induced the shape deformation of PEDOT:PSS, which reduced both sheet resistance and surface tension. When applied to the GO–Ag NW electrode, the PEDOT:PSS mixed with 7.5 wt % Triton X-100 provided the lowest sheet resistance. The optimal coating conditions for PEDOT:PSS mixed with 7.5 wt % Triton X-100 were 30 s of coating at 1000 rpm. The sheet resistance of the TP/GO–Ag NW electrode coated under these optimal conditions was 160 mΩ/sq., which was 49.73% and 82.10% lower than those of the GO–Ag NW and pure PEDOT:PSS coated GO–Ag NWs (P/GO–Ag NWs) electrodes, respectively. The driving performance of TP/GO–Ag-NW-based IEAPs was significantly better than that of the IEAPs based on GO–Ag NWs and P/GO–Ag NWs. Furthermore, the shape and network connection of the Ag NWs in the TP/GO–Ag-NW-based IEAPs was well maintained, as revealed by SEM images. Therefore, both the stability and durability of TP/GO–Ag-NW-based IEAPs were confirmed to improve. These results demonstrate the possibility of improving electrodes with high contact resistance in terms of durability and stability.

Author Contributions: Initial idea and writing—original draft preparation, M.P.; experiments and analysis of experiments data, S.Y.; investigation and data curation, Y.B.; experimental data curation and validation, S.K.; project administration and writing—review and editing, M.J.

Funding: This research was supported by a grant to the Bio-Mimetic Robot Research Center Funded by the Defense Acquisition Program Administration. It was also supported by the Agency for Defense Development in 2019 (UD160027ID) and the Basic Science Research Program through the National Research Foundation of Korea (NRF) funded by the Ministry of Education (NRF-2016R1D1A1B01011724).

Conflicts of Interest: The authors declare no conflict of interest.

References

1. Kim, J.; Jeon, J.H.; Kim, H.J.; Lim, H.; Oh, I.K. Durable and water-floatable ionic polymer actuator with hydrophobic and asymmetrically laser-scribed reduced graphene oxide paper electrodes. *ACS Nano* **2014**, *8*, 2986–2997. [CrossRef] [PubMed]
2. Kim, K.; Palmre, V.; Jeon, J.H.; Oh, I.K. IPMCs as EAPs: Materials. In *Electromechanically Active Polymers: A Concise Reference*; Springer: Cham, Switzerland, 2016; pp. 151–170.
3. Shahinpoor, M.; Kim, K.J. Ionic polymer-metal composites: I. Fundamentals. *Smart Mater. Struct.* **2001**, *10*, 819. [CrossRef]
4. Fujiwara, N.; Asaka, K.; Nishimura, Y.; Oguro, K.; Torikai, E. Preparation of gold–solid polymer electrolyte composites as electric stimuli-responsive materials. *Chem. Mater.* **2000**, *12*, 1750–1754. [CrossRef]
5. Palmre, V.; Lust, E.; Jänes, A.; Koel, M.; Peikolainen, A.L.; Torop, J.; Johanson, U.; Aabloo, A. Electroactive polymer actuators with carbon aerogel electrodes. *J. Mater. Chem.* **2011**, *21*, 2577–2583. [CrossRef]
6. Lee, D.Y.; Park, I.S.; Lee, M.H.; Kim, K.J.; Heo, S. Ionic polymer–metal composite bending actuator loaded with multi-walled carbon nanotubes. *Sens. Actuator A Phys.* **2007**, *133*, 117–127. [CrossRef]
7. Akle, B.J.; Leo, D.J. Single-walled carbon nanotubes—Ionic polymer electroactive hybrid transducers. *J. Intell. Mater. Syst. Struct.* **2008**, *19*, 905–915. [CrossRef]
8. Akle, B.; Nawshin, S.; Leo, D. Reliability of high strain ionomeric polymer transducers fabricated using the direct assembly process. *Smart Mater. Struct.* **2007**, *16*, S256. [CrossRef]
9. Palmre, V.; Brandell, D.; Mäeorg, U.; Torop, J.; Volobujeva, O.; Punning, A.; Johanson, U.; Kruusmaa, M.; Aabloo, A. Nanoporous carbon-based electrodes for high strain ionomeric bending actuators. *Smart Mater. Struct.* **2009**, *18*, 095028. [CrossRef]
10. Fukushima, T.; Asaka, K.; Kosaka, A.; Aida, T. Fully plastic actuator through layer-by-layer casting with ionic-liquid-based bucky gel. *Angew. Chem. Int. Ed.* **2005**, *44*, 2410–2413. [CrossRef] [PubMed]
11. Akle, B.J.; Bennett, M.D.; Leo, D.J.; Wiles, K.B.; McGrath, J.E. Direct assembly process: A novel fabrication technique for large strain ionic polymer transducers. *J. Mater. Sci.* **2007**, *42*, 7031–7041. [CrossRef]
12. Yoo, S.; Park, M.; Kim, S.; Chung, P.S.; Jeon, M. Graphene Oxide-Silver Nanowires Paper Electrodes with Poly (3,4-ethylenedioxythiophene)-Poly (styrenesulfonate) to Enhance the Driving Properties of Ionic Electroactive Polymer Actuators. *Nanosci. Nanotechnol. Lett.* **2018**, *10*, 1107–1112. [CrossRef]
13. Hwang, B.; Lim, S. PEDOT: PSS Overcoating Layer for Mechanically and Chemically Stable Ag Nanowire Flexible Transparent Electrode. *J. Nanomater.* **2017**, 1489186.
14. Yang, L.; Zhang, T.; Zhou, H.; Price, S.C.; Wiley, B.J.; You, W. Solution-processed flexible polymer solar cells with silver nanowire electrodes. *ACS Appl. Mater. Interface* **2011**, *3*, 4075–4084. [CrossRef] [PubMed]
15. Mayousse, C.; Celle, C.; Carella, A.; Simonato, J.P. Synthesis and purification of long copper nanowires. Application to high performance flexible transparent electrodes with and without PEDOT: PSS. *Nano Res.* **2014**, *7*, 315–324. [CrossRef]
16. Park, M.; Kim, J.; Song, H.; Kim, S.; Jeon, M. Fast and Stable Ionic Electroactive Polymer Actuators with PEDOT: PSS/(Graphene–Ag-Nanowires) Nanocomposite Electrodes. *Sensors* **2018**, *18*, 3126. [CrossRef] [PubMed]
17. Koley, D.; Bard, A.J. Triton X-100 concentration effects on membrane permeability of a single HeLa cell by scanning electrochemical microscopy (SECM). *Proc. Natl. Acad. Sci. USA* **2010**, *107*, 16783–16787. [CrossRef] [PubMed]
18. Tevi, T.; Saint Birch, S.W.; Thomas, S.W.; Takshi, A. Effect of Triton X-100 on the double layer capacitance and conductivity of poly (3,4-ethylenedioxythiophene): Poly (styrenesulfonate)(PEDOT: PSS) films. *Synth. Met.* **2014**, *191*, 59–65. [CrossRef]

19. JeongáLee, S. Improved stability of transparent PEDOT: PSS/Ag nanowire hybrid electrodes by using non-ionic surfactants. *Chem. Commun.* **2017**, *53*, 8292–8295.

20. Oh, J.Y.; Shin, M.; Lee, J.B.; Ahn, J.H.; Baik, H.K.; Jeong, U. Effect of PEDOT nanofibril networks on the conductivity, flexibility, and coatability of PEDOT: PSS films. *ACS Appl. Mater. Interface* **2014**, *6*, 6954–6961. [CrossRef] [PubMed]

21. Hummers, W.S., Jr.; Offeman, R.E. Preparation of graphitic oxide. *J. Am. Chem. Soc.* **1958**, *80*, 1339. [CrossRef]

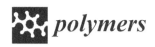

Article

Ultrasensitive Wearable Strain Sensors of 3D Printing Tough and Conductive Hydrogels

Jilong Wang [1], Yan Liu [1], Siheng Su [2], Junhua Wei [3], Syed Ehsanur Rahman [3], Fuda Ning [4], Gordon Christopher [3], Weilong Cong [5] and Jingjing Qiu [3,*]

1   Key Laboratory of Textile Science & Technology of Ministry of Education, College of Textiles, Donghua University, Shanghai 201620, China; jilong.wang@dhu.edu.cn (J.W.); yan.liu@mail.dhu.edu.cn (Y.L.)
2   Department of Mechanical Engineering, California State University Fullerton, Fullerton, CA 92831, USA; ssu@fullerton.edu
3   Department of Mechanical Engineering, Texas Tech University, 2500 Broadway, Lubbock, TX 79409, USA; junhua5wei@gmail.com (J.W.); syed.rahman@anton-paar.com (S.E.R.); gordon.christopher@ttu.edu (G.C.)
4   Department of Systems Science and Industrial Engineering, State University of New York at Binghamton, Binghamton, NY 13902, USA; fning@binghamton.edu
5   Department of Industrial Engineering, Texas Tech University, 2500 Broadway, Lubbock, TX 79409, USA; weilong.cong@ttu.edu
*   Correspondence: jenny.qiu@ttu.edu

Received: 15 September 2019; Accepted: 11 November 2019; Published: 13 November 2019

Abstract: In this study, tough and conductive hydrogels were printed by 3D printing method. The combination of thermo-responsive agar and ionic-responsive alginate can highly improve the shape fidelity. With addition of agar, ink viscosity was enhanced, further improving its rheological characteristics for a precise printing. After printing, the printed construct was cured via free radical polymerization, and alginate was crosslinked by calcium ions. Most importantly, with calcium crosslinking of alginate, mechanical properties of 3D printed hydrogels are greatly improved. Furthermore, these 3D printed hydrogels can serve as ionic conductors, because hydrogels contain large amounts of water that dissolve excess calcium ions. A wearable resistive strain sensor that can quickly and precisely detect human motions like finger bending was fabricated by a 3D printed hydrogel film. These results demonstrate that the conductive, transparent, and stretchable hydrogels are promising candidates as soft wearable electronics for healthcare, robotics and entertainment.

Keywords: hydrogels; 3D printing; tough; sensor

---

1. Introduction

As we know, human skin is soft, self-healable and stretchable, and has the ability to sense subtle external changes. This amazing property has attracted tremendous interest in artificial skin, especially wearable electronics for healthcare, artificial intelligence and soft robotics [1–3]. These artificial skin-like devices can monitor environmental stimuli such as pressure, strain, temperature, and deformation by detecting electric signals like current and voltage, or measuring electrical properties including resistance, and capacitance. "Electronic skin" is usually considered as a stretchable sheet with area above 10 cm^2 integrating sensors to detect different external stimuli [4]. Usually, electronic skin is made of stretchable electrical conductors including carbon grease [5], graphene sheets [6], carbon nanotubes [7], liquid metals [8] and metal nanostructures [9,10], which transmits signals via electrons. Although these materials present high conductivity and excellent stretchability, which meet the necessary requirements of electronics skin, they fail to meet other additional requirements like biocompatibility and transparency. On the other side, human skin can report signals via ions, which provides a potential pathway to develop ionic conductors based on a sensory sheet called "ionic skin" [4]. Hydrogels are three-dimensional networks composed of high-molecular weight polymer, large amounts of water, and crosslinkers [11]. As the water in hydrogels can dissolve ions, hydrogels can be employed as ionic conductors [4,12], which may have potential applications in "ionic skin". In addition, hydrogels are highly stretchable and biocompatible [13]. Furthermore, high transparency of hydrogels allows these sensory sheets to report electrical signals without impeding optical signals [4]. They can behave as tough as elastomers due to recent developments [14–17], which can monitor large deformation, like finger bending.

Three-dimensional (3D) printing, also known as an additive manufacturing process, is an emerging technology [18]. Due to its rapid production with high shape fidelity, 3D printing technology has attracted tremendous attention since it was first proposed by Charles W. Hull in 1986 [19]. Recently, volumetric additive manufacturing has been developed and has received lots of attention due to its excellent performance to overcome limitations of low speed and geometric constraints [20–22]. However, the 3D printing technology has been recently introduced to fabricate hydrogels. Extrusion printing method is a modified fused deposition modeling method that extrudes continuous liquid inks to achieve layered structures. As the extrusion printing method has lots of advantages including simple fabrication procedure, large range of materials, good balance between printer's cost and printing quality, and high cell deposition in bioprinting, it has been considered as an excellent choice to print hydrogels [17,23].

Various polymers have relatively high viscosity to maintain their pattern in printing process, and have crosslinking abilities allowing for 3D structures maintenance after printing, like collagen [24], hyaluronic acid (HA) [25], chitosan [26] and alginate [18], have been employed in 3D printing technology to achieve 3D printed hydrogels. Usually, physical crosslinking can be induced by temperature change [15,24] and ionic crosslinking [11,27], whereas chemical crosslinking can be formed by polymerization [28]. Sodium alginate (SA), an anionic polymer isolated from brown algae, has the ability to crosslink assisting by divalent or trivalent ions [11]. Due to its high biocompatibility, hydrophilicity and biodegradability under normal physiological conditions, SA has received increasing attention as an instant gel for tissue engineering. Although there are lots of conventional methods to fabricate SA hydrogel constructs including the injection molding method and solution casting method, SA solution has a certain viscosity and its limited flowability leads to a poor dispersion in molds [11]. Compared to these conventional methods, 3D printing technology has one main advantage to fabricate customized constructs, which can fit various requirements of wearable sensors applied on different body parts on different humans [17]. In addition, 3D printing technology has the potential to fabricate hydrogels with hierarchically porous structures or gradient properties, which may improve sensitivity and sensing range of wearable sensors [29]. Although 3D printing technology has these advantages to fabricate SA hydrogel constructs, several challenges have not been well addressed, which limits its development. One of the common challenges of 3D printing hydrogels is to achieve printed constructs with high shape fidelity due to low viscosity leading to a collapse tendency of the printed constructs. Various methods have been proposed to increase the viscosity of SA solution, such as increasing the SA ink concentration or varying molecular weight [30], combining with other materials including nanocellulose [18] or gelatin [31], employing a supporting sacrificial polymer [32] and partially crosslinking alginate with calcium ions [33].

To improve the printing resolution and mechanical property, the hybrid agar/calcium alginate (CA)/polyacrylamide(PAAm) hydrogels combining brittle thermo-responsive agar and ionic-responsive alginate and soft polyacrylamide network is proposed in this manuscript. During printing, mixture of thermo-responsive agar and ionic-responsive alginate was extruded as a continuous stripe to enhance printing resolution. Meanwhile, due to the increasing viscosity, the mixture can maintain its shape during printing. After photopolymerization and solution soaking, the 3D printed tough hydrogels were achieved. With calcium crosslinking of alginate, tensile strength and fracture energy of 3D printed hydrogels are greatly improved. Furthermore, the water in 3D printed hydrogels dissolves lots of calcium ions, which make them work as ionic conductors. A wearable soft resistive strain sensor was developed by a 3D printed hydrogel film. This resistive strain sensor exhibits quick and accurate detection of changes of finger bending, which demonstrates that the conductive, transparent, stretchable hydrogels can be used as wearable resistive strain sensor to monitor human motion.

2. Experimental Section

2.1. Materials

Sodium alginate was received from FMC biopolymer (Rockland, ME, USA). Agar, acrylamide, N,N'-methylenebis (acrylamide) (MBAA), Irgacure 2959 and calcium chloride were ordered from Sigma-Aldrich (St. Louis, MO, USA) without further purification. An acrylic elastomer (VHB 4905) was received from 3M (St. Paul, MN, USA). Copper tape was used to connect conductive gels and electric wires.

2.2. 3D Printing System

As shown in Figure 1, a modified Leapfrog 3D printer was employed to fabricate tough hydrogels as described in our previous literature [18]. In brief, a syringe pump (NE-500 OEM, New Era, Gawler, Australia) was installed onto 3D printer to extrude pre-gel ink at a controlled infusion velocity ($0.73~\mu L~h^{-1}$–2100 mL h^{-1}), and a thermal pad (HEATER-KIT-5SP, New Era, Gawler, Australia) was used to wrap up syringe to maintain printing temperature. The blunt tip needles (gauge 14–26) were used to inject continuous hydrogel solution. A commercial software "Simplify 3D" was applied to control printing process.

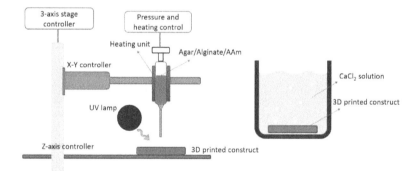

Figure 1. Schematic diagram of a 3D printing procedure.

2.3. 3D Printing Hydrogels Fabrication

The 3D printing procedure was performed on a modified Leapfrog 3D printer that is similar to our previous literature [17]. SA 200 mg was first dissolved in 10 mL DI water with continuous stirring overnight. Then the SA solution was heated to 95 °C in an oil bath, and 200 mg of agar was added into the solution. After agar was fully dissolved in the water and 3000 mg acrylamide and corresponding MBAA, Irgacure 2959 were added, the hybrid ink was cooled to 55 °C and ready for printing. As shown in Figure S1, three different infill angles were used to achieve 3D printing constructs. The width and length of design were set at 100 mm. The thickness of 3D printing constructs was around 1 mm. After printing, the printed construct was exposed to UV light (365 nm) for 1 h. Then 100 mM CaCl$_2$ solution was used to crosslink sodium alginate for 15 min. To achieve conductive printed hydrogels, the printed gels was soaking in 100 mL 100 mM CaCl$_2$ solution overnight to increase the concentration of Ca^{2+} ions in printed gels. These excess Ca^{2+} ions can be used as ions carriers. This sample was labelled as A2C2. The formula and labels of other samples were summarized in Table 1.

Table 1. The formula of printing ink in 10 mL Deionized (DI) water.

Sample	Agar (mg)	Sodium Alginate (mg)	Irgacure 2959 (mg)	Acrylamide (mg)	MBAA (mg)	Concentration of CaCl$_2$ (mM)
A1C2	100	200	90	3000	3	100
A2C2	200	200	90	3000	3	100
A3C2	300	200	90	3000	3	100
A2C1	200	100	90	3000	3	100
A2C3	200	300	90	3000	3	100
A1S2	100	200	90	3000	3	N/A
A2S2	200	200	90	3000	3	N/A

2.4. Mechanical Test

The tensile test and pure shear test were performed on a universal tensile machine (AGS-X, SHIMADZU, Kyoto, Japan) at room temperature. Each sample was measured in triplicate. The tensile measurement was performed at a crosshead speed of 10 mm min^{-1}. The stress with 0 to 10% strain was used to calculate the elastic modulus (E). The stress σ was calculated by the following equation [34].

$$\sigma = \frac{F}{WT} \tag{1}$$

where F is force, W and T mean width and thickness of sample. The pure shear test is used to calculate the fracture energy. Two same samples are used at one test, one notched sample and one un-notched sample, and the notched sample is measured at first to get the point at which crack propagation began, while the un-notched sample is measured to get the force-displacement curve. The fracture energy is calculated by the following equation [35].

$$Fracture\ energy = \frac{\int_{l_0}^{l_c} F dl}{WT} \tag{2}$$

where W and T mean width and thickness of sample, l_c represents critical distance, at which point crack propagation occurs, lo means initial length of sample.

2.5. Rheological Test

The rheological properties of the SA solution with various compositions were analyzed using an AR-G2 stress-controlled rheometer (TA Instrument, New Castle, DE, USA) with a parallel plate geometry (1 mm) at 45 °C. The viscosity of hybrid ink was also measured with different temperatures from 35 °C to 65 °C. The shear rate was varied from 0.01 s^{-1} to 1000 s^{-1}. The oscillation frequency measurements were conducted at 35 °C and at a frequency of 1 Hz to measure storage and loss modulus of ink solution. The tan δ values was calculated as [18]:

$$Tan\ \delta = \frac{G''}{G'} \tag{3}$$

where G' and G'' are storage and loss modulus, respectively.

2.6. Conductivity Test

The hydrogel sample was clipped onto a universal tensile machine (AGS-X, SHIMADZU, Kyoto, Japan) at room temperature around 20 °C. The speed was 10 mm min^{-1}. After the hydrogel sample was stretched at certain strain, the resistance was measured by a multi-meter. To limit water evaporation from 3D printed hydrogels, VHB were used to wrap up hydrogel sample during test. To achieve a wearable strain sensor, 3D printed hydrogels covered with VHB were connected to electric wires and placed on a finger with the help of copper tape. To study conductivity of hydrogels at various concentration of ions and SA, calcium alginate (CA)/polyacrylamide(PAAm) hydrogels were fabricated via an injection molding method as previous work [11]. Different concentration of CaCl$_2$ (50 mM, 100 mM, 300 mM and 500 mM) were used to soak hydrogels. Three concentrations of alginate (10 mg/mL, 20 mg/mL and 30 mg/mL) and acrylamide (1.69 M, 2.53 M and 3.38 M) were used.

2.7. Cytotoxicity Test

The cytotoxicity test of hydrogels was performed as previous method using U87-MG glioblastoma cells (ATCC HTB-14) cultured by Eagle's Minimum Essential Medium (EMEM) with 10% fetal bovine serum (FBS) and 100 IU ml^{-1} penicillin and streptomycin [13,36]. The hydrogels were soaked in DI water for 8 days, followed by being immersed in the cell cultural medium for 3 days to remove residual monomer and chemicals. After that, the washed hydrogels were put into another cell cultural medium again for 3 days to achieve conditioned cell cultural medium. Following this, 2×10^4 U87-MG cells were seeded onto a 24-well plate and cultured in fresh cell cultural for 3 days. Then the conditioned cell cultural medium was used to culture the cells for another 24 h or 48 h. To evaluate the cytotoxicity of hydrogels, a crystal violet staining method was used. First, cells were fixed by 1 mL of 2% glutaraldehyde in triplicate for 10 min. Then the medium was aspirated and the cells were washed with phosphate-buffered saline (PBS) twice. Then, 1 mL of 0.1% crystal violet was added into the plate and the cells were stained for 40 min. Crystal violet was then washed away by water. Crystal violet was present only in the stained cells. The plate was further drained inversely overnight. The cytotoxicity was evaluated by reading the absorbance at 590 nm of the crystal violet.

In addition, Trypan blue was used to show the cell viability by the previous method [13,36]. Specifically, old media in the 24 well plate were aspirated and 1 mL of Dulbecco's phosphate-buffered saline (DPBS) was added to wash the adherent cells. Then, DPBS was added into the counting chamber, and the chamber was inserted into a Cellometer Vision Image Cytometry for counting and imaging.

2.8. Swelling Test

The swelling test was performed by soaking four hydrogel samples (400 mg) in DI water for different time periods. The swelling ratio (SR) was defined as [13]:

$$SR = \frac{Ws - Wo}{Wo} \times 100\% \tag{4}$$

where *Ws* and *Wo* represent the weight of hydrogels after swelling in DI water in different time periods and the weight of hydrogels before swelling, respectively.

2.9. Statistical Analysis

A one-way analysis of variance (ANOVA) with Fisher's pair-wise multiple comparison was employed to analyze the data. A *P*-value smaller than 0.05 was considered statistically significant.

3. Results and Discussions

3.1. Viscosity and Printability

To achieve 3D printable tough and conductive hydrogels, pre-gel solution was prepared by mixing sodium alginate (SA), agar (Ag), acrylamide (AAm), *N,N'*-methylenebis (acrylamide) (MBAA) and Irgacure 2959 into DI water. To well understand printability of hydrogels, the viscosity of pre-gel solution was well characterized by an AR-G2 stress-controlled rheometer. As shown in Figure 2a, the pure SA solution shows a low zero-shear viscosity that is below 30 mPa s that is regarded as minimum solution viscosity for extrusion printing in previous literature, leading to poor shape fidelity [23]. The increased viscosity can help to improve shape fidelity during printing. To improve the shape fidelity of SA, thermo-responsive agar gel solution was added into the 3D printing ink formulation. Compared to SA, agar is a thermo-responsive polymer, which exhibits high viscosity at low temperature. The mixture agar and SA exhibited an extremely higher viscosity at low shear rate and was shear thinning. With temperature decreasing, the viscosity of mixture dramatically increased because of agar gelation (the insert in Figure 2a). After adding a large amount of acrylamide, the printing ink was prepared, and the change of viscous behavior was negligible. Figure 2b showed storage and loss modulus, and tan δ of ink solution (SA 200 mg, Ag 200 mg, and AAm 3000 mg). Tan δ is the ratio of loss modulus to storage modulus. The printing ink solution had a tan δ value below 1, indicating this printing ink is more gel-like than liquid [16].

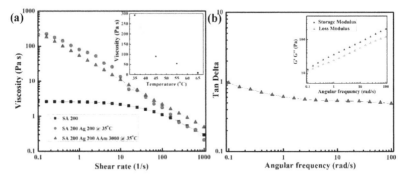

Figure 2. (a) Rheological data for ink formulation, the insert: Viscosity of Sodium alginate (SA) 200/Ag 200 with different temperature at shear rate 1/s, and (b) Tan δ of the ink solution, the inset: the storage and loss modulus of ink solution.

3.2. Mechanical Properties

To well investigate mechanical properties of 3D printed hydrogels, tensile test and pure shear test were performed to characterize strength and fracture energy of 3D printed hydrogels. After photopolymerization and calcium chloride soaking, 3D printed tough hydrogels were successfully achieved. As shown in Figure 3a, the 3D printed hydrogels could be easily wrapped into a knot without any damage. After stretching, this knot could return to the original state. This result indicates that the 3D printed tough hydrogels own excellent mechanical properties and good elasticity. The stress-strain curves showed that the 3D printed hydrogels without calcium chloride solution soaking own a weak tensile strength, however, the tensile strength is largely improved after soaking process (as shown in Figure 3b). The improved tensile strength derives from formation of calcium ionically crosslinked alginate network during soaking process. The broken of hydrogels usually contains two sequential steps: initial fracture formation (nucleation) and subsequent fracture propagation (growth) [17]. In A2S2 gels, no nucleation is found due to the agar chain pullout mechanism [37]. With increasing strains, the agar chains pullout from the aggregated agar helical bundles progressively. However, the agar network remains integrated during this process. With addition of alginate chains, the weak entanglements between chains nearly restricts the pullout process of agar. This mechanism allows that the A2S2 gels own relatively long elongation. However, the A2C2 gels showed a distinctive breaking process with a higher strength and smaller elongation. With the addition of the Ca^{2+}, the alginate chains are ionically crosslinked. The stress is concentrated on the ionically crosslinked alginate chains and unzip ionically crosslinked alginate chains preferentially leading to breakage of printed hydrogels at relatively low elongation. Therefore, A2C2 gels achieved the higher toughness at 603.22 ± 61.78 kJ m^{-3} while the A2S2 gels depicted a little smaller toughness at 493.27 ± 42.00 kJ m^{-3}, although the A2C2 had a larger strength 488.75 ± 58.31 kPa and the A2S2 owned a very small strength at 142.67 ± 19.60 kPa (Table 2). In addition, the mechanical properties of printed gels with different compositions have been also summarized in Table 1 and Figure S2. With increasing amount of alginate and agar, the strength and toughness are improved, however, the elongation is nearly changed due to the existence of calcium crosslinked alginate network. Table S2 and Figure 3c summarized the mechanical properties of the printed gels with various printing infill angles. Three different infill angels including 0°, 45° and 90° were employed in this experiment and the design of 3D printing constructs was presented in Figure S1. The mechanical strength and elongation of these printed gels were similar. Therefore, the presence of calcium crosslinking of alginate network makes the infill have no influence on mechanical properties of printed gels.

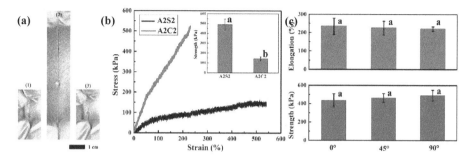

Figure 3. (**a**) Photograph of 3D tough hydrogels, (1) original state, (2) under tension and (3) back to original state, (**b**) stress-strain curve of 3D printed gels with or without CaCl$_2$ soaking, the insert is strength of 3D printed gels, means with different letters are statistically different at P < 0.05, (**c**) strength and elongation of 3D printed gels with different printing angles (A2C2), (P > 0.05), means with different letters are statistically different at P < 0.05.

Table 2. The mechanical properties of the printed hydrogels with different composition.

Sample	Young's Modulus (kPa)	Strength (kPa)	Elongation (%)	Toughness (kJ m^{-3})
A1C2	26.74 ± 4.95	385.56 ± 31.10	223.63 ± 62.96	479.71 ± 150.97
A2C2	38.14 ± 2.99	488.75 ± 58.31	220.30 ± 11.74	603.22 ± 61.78
A3C2	55.45 ± 9.33	744.57 ± 144.17	235.78 ± 31.95	1049.77 ± 241.21
A2C1	17.14 ± 4.63	283.33 ± 31.07	215.44 ± 12.10	362.01 ± 31.26
A2C3	41.20 ± 3.61	596.84 ± 28.43	218.54 ± 24.81	762.02 ± 43.76
A2S2	16.29 ± 0.54	142.67 ± 19.60	566.38 ± 23.47	493.27 ± 42.00

The pure shear test was used to calculate the fracture energy of 3D printed tough hydrogels. The schematic diagram of the pure-shear test was presented in Figure 4a and detail of measurement was located in experimental section. The fitted stress-strain curves of notched and un-notched hydrogels clearly were presented in Figure S3. According to Equation (2), the fracture energy of 3D printed hydrogels with different compositions were calculated. As shown in Figure 4b, with calcium ionically crosslinked alginate, the fracture energy is largely improved due to increased crosslinking points. When increasing the amount of agar chains, the fracture energy also increases, which demonstrates that higher crosslinking degree leads to higher fracture energy.

As shown in Figure 5a, the fitted stress-strain curve of notched samples with various compositions clearly presents that the strength at fracture are both improved with increasing amount of agar and with calcium chloride solution soaking. In addition, we can clearly find that the weak gel (A1C2 and A1S2) crack propagates very quickly and even immediately (Figure 5a,b). That means the notched sample is totally treated in a quick time. However, with increasing agar amount, the crosslinking degree is improved and the notched gel sample of A2S2 is slowly torn (Figure 5c). However, after calcium chloride soaking, the 3D printed tough gels (A2C2) have much slower crack propagation (Figure 5d). That means the combination of agar and calcium alginate (CA) network can greatly prevent the crack propagation.

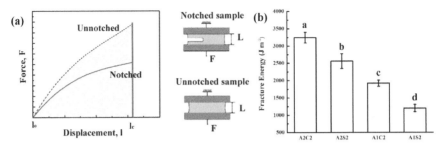

Figure 4. (**a**) Schematic diagram of the pure-shear test for measuring fracture energy of hydrogels, and (**b**) Fracture energy of 3D printed hydrogels, means with different letters are statistically different at P < 0.05.

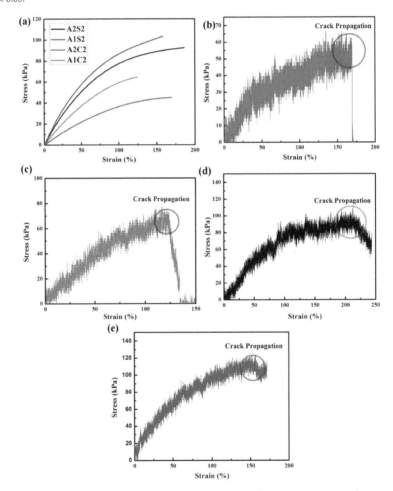

Figure 5. (**a**) Fitted stress-strain curve of notched samples with different compositions, and stress-strain curve of notched sample (**b**) A1S2, (**c**) A1C2, (**d**) A2S2, and (**e**) A2C2.

3.3. Swelling and Cytotoxicity

The swelling properties of printed gels are also systematically investigated. In the soaking process, the water molecules penetrate into polymeric hydrogels resulting in an expansion of polymeric networks and a low concentration region of polymeric chains, which leads to mechanical fracture [11]. Therefore, different chemical structure and crosslinking density can result in different swelling properties. As shown in Figure 6a, the A2S2 gels depict a much larger swelling ability, compared to A2C2 printed gels. The presence of calcium ionically crosslinked alginate not only increases the crosslinking degree, but also restricts the agar and polyacrylamide chains, causing limited swelling ability.

On the other hand, with increasing amount of agar and calcium crosslinked alginate, the swelling ability is quenched, which demonstrates that the high amount of hard network leads to an inferior swelling ability (Figure S4a and S4b). These results are consistent to the mechanical properties of printed gels. As shown in Figure S4c, the printed gels with different infill angles present exactly the same swelling ability, which demonstrates that the infill angles have no influence on chemical structures and crosslinking degree.

The cytotoxicity of 3D printed hydrogels was also evaluated via U87-MG cells. Figure 6a depicted similar viable cell quantities between the control group and A2S2 hydrogels conditioned group in both 24 and 48 h. These results reveal that the 3D printed hydrogels own high biocompatibility after removing the unreacted acrylamide monomer. Compared to 24 h culture group, the number of viable cells after 48 h culture increased. This result demonstrates that the gels conditioned medium cannot affect cellular reproduction. The Figure S5 presented the live and dead cell images. Cells circled with green were viable, indicating these 3D printed hydrogels own high biocompatibility. After removal of the unreacted acrylamide monomer and other residues, the 3D printed hydrogels own low cytotoxicity with excellent mechanical strength and toughness, which can be considered as a potential candidate in wearable electronics.

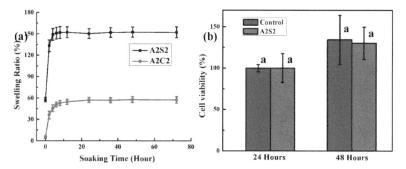

Figure 6. (**a**) Swelling ratio of A2C2 and A2S2 gel, (**b**) Cell viability of U87-MG cells after culturing 24 h and 48 h ($P > 0.05$), means with different letters are statistically different at $P < 0.05$.

3.4. Conductivity and Sensors

As shown in Figure S6a, high transparency of 3D printed gel was achieved, which allowed these 3D printed hydrogels to detect resistance without affecting optical signals. Due to existence of large amounts of water inside hydrogels, hydrogels offer physical similarity to biotissues, and also own excellent capability to contain lots of ions [38]. By introducing Ca^{2+} and Cl^- ions in the 3D printed hydrogels, conductive hydrogels could be used as ionic wires in the circuit. The Figure 7a depicted the bright light when 3D printed conductive hydrogels were connected into the circuit. The light was off, when conductive hydrogels was moved out of circuit. It demonstrates that 3D printed hydrogels are ionically conductive by containing Ca^{2+} and Cl^- ions. The resistance of conductive hydrogels is measured via a multi meter. The conductivity of 3D printed hydrogels is around 13.9 mS/cm, which is similar to the conductivity of electrolyte, which demonstrates that the conductivity of conductive hydrogels is close to the conductivity of calcium chloride solution. To well investigate conductivity of hydrogels, calcium alginate (CA)/PAAm double network (DN) hydrogels were fabricated via injection molding method according to our previous work with various concentration of ions, SA and AAm [11]. As shown in Figure S7a, with increasing concentration of $CaCl_2$, the conductivity of DN hydrogels enhances. In addition, the conductivity of DN hydrogels with various concentration of $CaCl_2$ were close to the that of $CaCl_2$ solution with same concentration. These results strongly indicate that conductivity of hydrogels derives from ions. Furthermore, Figure S7b and S7c exhibited no significant change of DN hydrogels' conductivity with various concentration of SA and AAm. It demonstrates that the polymeric network in DN hydrogels has no effect on conductivity of ionically conductive hydrogels.

Figure 7b showed reduced brightness of the light when the 3D printed hydrogels were stretched, indicating that the stretch largely enhanced the resistance of hydrogels. This result exhibits that the conductivity of these conductive hydrogels is dependent on the strains. When the stretch was released, the brightness of light turned up again. To further investigate the resistance change of 3D printed hydrogels, the resistance change was measured with various stretches. We first assume the 3D printed hydrogels are incompressible and the conductivity is constant during stretching (Figure 7c). The resistance ratio is given by $R/R_0 = (L/L_0)^2$, where R and R_0 mean the resistance of the stretched hydrogels and initial hydrogels, respectively. Figure 7d showed that the experimental data of the conductive hydrogels were close to the curve of theoretic equation. The small deviation of experimental data from theoretic curve might be caused by damage in the hydrogels. These results indicate that the measured resistance of the conductive hydrogels is reasonable. By comparison with other electronic conductors like indium tin oxide (ITO), silver nanowires (AgNWs), graphene, single-wall nanotubes (SWNTs), the conductive hydrogels, ionic conductors, owned lower conductivity than these mentioned electronic conductors. However, when high transmittance and stretchability are necessary, these conductive, transparent and stretchable hydrogels have specific advantages. At high stretch, these hydrogels had lower sheet resistance than these above mentioned electronic conductors. This result is consistent with previous literature [39].

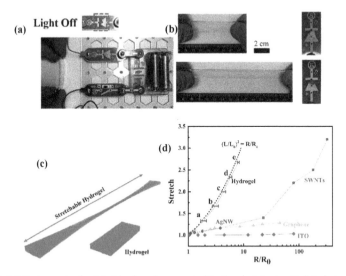

Figure 7. (**a**) Photograph of printed hydrogels connected in the electric circuit, (**b**) photographs of the light changes of elongation connected in the electric circuit, (**c**) schematic diagram of hydrogels when stretching, (**d**) the normalized resistance of printed hydrogels is measured as a function of stretch, means with different letters are statistically different at P < 0.05, plotted against the ideal geometric behavior, and normalized resistance for ITO [40], AgNWs [41], graphene [42], SWNTs [43].

As depicted in Figure 8a, the resistant change of these conductive hydrogels showed a good linear correlation to strain in a range of 0–1.5, which indicates that the conductive hydrogels own a relatively large sensing range. The strain sensitivity of conductive hydrogels can be defined as the slope of resistance change rate ($R-R_0/R_0=\Delta R/R_0$) versus applied strain (λ), formulized as $S=\delta(\Delta R/R_0)/\delta\lambda$ [38]. A gauge factor (3.83) was achieved via these conductive hydrogels that is superior to previously reported hydrogel-based strain sensor (0.478) [38]. This result demonstrates the conductive hydrogels exhibit a high sensing sensitivity. As shown in Figure 8b, the relative resistance change of these conductive hydrogels was exhibited during a step-by-step loading-unloading cycle at different strains. This curve clearly showed the relative resistance of conductive hydrogels had a step-like trend. In addition, the relative resistance directly increased or decreased when conductive hydrogels were stretched or released to certain strain. It is meaningful that conductive hydrogels have quick response ability, because no hysteresis was found during strain change. Moreover, a good sensing stability of conductive hydrogels was also observed, since the relative resistance was kept stable during load-holding or unload-holding period at different strain. As a wearable strain sensor, it is also important for the conductive hydrogels to own high stability. The Figure 8c showed the relative resistance change for 100 tensile cycles under 10% strain. The resistance of the conductive hydrogels was similar to the original level and these hydrogels showed no visible damage or delamination after 100 cycles of 10% strain. These results demonstrate that the conductive hydrogels have superior stability in sensing. The Figure 8d showed no obvious relative resistance change in temperature range of 20 to 40 °C, which means that these conductive hydrogels own high reliability when they are attached onto the human body.

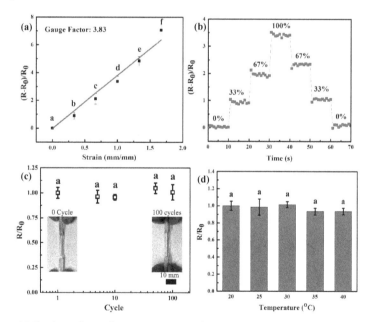

Figure 8. (a) The dependence of sensing sensitivity of conductive hydrogels with the applied strain. The strain sensitivity (S) can be defined as the slope of resistance change rate ($\Delta R/R_0$) versus applied strain (λ), formulized as $S = \delta(\Delta R/R_0)/\delta\lambda$, means with different letters are statistically different at $P < 0.05$, (b) The relative resistance changes vs time when a loading–unloading cycle of conductive hydrogels at different strains, (c) the resistance ratio of conductive hydrogels as a function of fatigue cycle number ($P > 0.05$), means with different letters are statistically different at $P < 0.05$, and (d) the resistance ratio of conductive hydrogels as a function of temperature ($P > 0.05$), means with different letters are statistically different at $P < 0.05$.

In addition, we developed a wearable resistive strain sensor by these 3D printed conductive hydrogels and fixed it onto an index finger by copper tapes to monitor finger bending (Figure 9a). When the index finger bended step-by-step, the relative resistance change of this sensor rose up in a step-like trend (Figure 9b), which is similar to results in Figure 8b. The relative resistance of this wearable resistive strain sensor directly enhanced without hysteresis after the finger bending to a gesture, which exhibits fast response ability. During gesture-holding process of the index finger, the resistance of the sensor could remain at a constant, which shows the good sensing stability of this wearable resistive strain sensor. The Figure 9c also depicted a repeatable response during finger bending, which demonstrated that this wearable strain sensor has a good sensing stability in repeated usage. These results indicate that this conductive hydrogels-based strain sensor can be applied for human motion monitoring with high sensing sensitivity and stability.

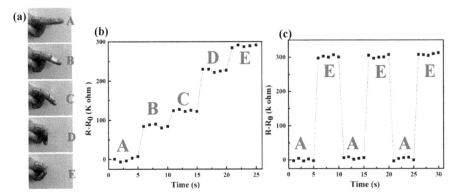

Figure 9. (**a**) Photographs of finger bending, (**b**) resistance change when finger bending and (**c**) repeated response of the resistive strain sensor.

4. Conclusion

In summary, a tough and conductive hydrogel was developed by 3D printing technology. The combination of alginate and agar guarantee high ink viscosity resulting in high printing precision. A double network structure that combined covalent crosslinking and Ca^{2+}-alginate coordination was employed to achieve conductive, transparent and stretchable hydrogels. The results demonstrated that the double network structure affords a smooth stress-transfer and recoverable energy dissipation to gift the hydrogels with superior mechanical strength and toughness. In addition, the 3D printed hydrogels containing a large amount of water that dissolves calcium ions, could work as ionic conductors. Conductive hydrogels depict quick, steady and repeated deformation toward strain to change the ionic transport, leading to rapid sensing response, high sensing stability and strain sensitivity. A wearable resistive strain sensor was fabricated connecting one 3D printed hydrogel film with conductive tape, which can rapidly and precisely detect the joint motions of finger bending. These results support the possibility for developing conductive, transparent, and stretchable hydrogels as wearable resistive strain sensor for human motion detection or sensory skin employed in a soft robot.

Supplementary Materials: The following are available online at http://www.mdpi.com/2073-4360/11/11/1873/s1, Figure S1: Design of 3D printed hydrogels with different infill angles (a) 0°, (b) 45° and (c) 90°, Figure S2: (a) Stress-Stain curve of 3D printed gel with different alginate concentration, (b) strength of 3D printed gels with different agar content (P > 0.05), means with different letters are statistically different at P < 0.05, (c) Stress-stain curve of 3D printed gel with different agar concentration, and (d) strength of 3D printed gels with alginate content (P > 0.05), means with different letters are statistically different at P < 0.05, Figure S3: The fitted stress-strain curve of both notched and unnotched sample (a) A1S2, (b) A1C2, (c) A2S2, and (d) A2C2, Figure S4: (a) swelling ratio of gel with different agar content, (b) swelling ratio of gel with different alginate content, and (c) Swelling ratio of A2C2 gel with different infill method, Figure S5: Live and dead cell image (a) control, and (b) A2S2, Figure S6: High transparency and conductivity of 3D printed hydrogels, Figure S7: (a) Conductivity of hydrogels (sodium

Polymers **2019**, *11*, 1873

alginate (SA) 200 mg and acrylamide (AAm) 1200 mg) by injection molding method with various concentration of calcium chloride, means with different letters are statistically different at P < 0.05, (b) Conductivity of hydrogels (AAm 1200 mg and CaCl2 100 mM) by injection molding method various alginate content, (P > 0.05), means with different letters are statistically different at P < 0.05, and (c) Conductivity of hydrogels (SA 200 mg and CaCl2 100 mM) by injection molding method various concentration of acrylamide, (P > 0.05), means with different letters are statistically different at P < 0.05, Table S1: the formula of printing ink in 10 mL DI water, Table S2: the mechanical properties of printed hydrogels with different printing parameter.

Author Contributions: Data curation, S.E.R. and F.N.; Formal analysis, J.W., S.E.R., F.N., G.C. and W.C.; Investigation, J.W., Y.L. and S.S.; Methodology, J.W.; Project administration, J.Q.; Resources, G.C., W.C. and J.Q.; Writing—original draft, J.W.; Writing—review & editing, J.W., Y.L., S.S. and J.W.

Funding: Fundamental Research Funds for the Central Universities: 19D110112

Acknowledgments: The authors thank the financial support from the Fundamental Research Funds for the Central Universities (19D110112). This research is also supported by the Initial Research Funds for Young Teachers of College of Textiles, Donghua University.

Conflicts of Interest: The authors declare no conflict of interest.

References

1. Hammock, M.L.; Chortos, A.; Tee, B.C.K.; Tok, J.B.H.; Bao, Z.A. 25th Anniversary Article: The Evolution of Electronic Skin (E-Skin): A Brief History, Design Considerations, and Recent Progress. *Adv. Mater.* **2013**, *25*, 5997–6037. [CrossRef]
2. Chortos, A.; Bao, Z.N. Skin-inspired electronic devices. *Mater. Today* **2014**, *17*, 321–331. [CrossRef]
3. Wang, X.D.; Dong, L.; Zhang, H.L.; Yu, R.M.; Pan, C.F.; Wang, Z.L. Recent Progress in Electronic Skin. *Adv. Sci.* **2015**, *2*, 1500169. [CrossRef]
4. Sun, J.Y.; Keplinger, C.; Whitesides, G.M.; Suo, Z.G. Ionic Skin. *Adv. Mater.* **2014**, *26*, 7608–7614. [CrossRef] [PubMed]
5. Pelrine, R.; Kornbluh, R.; Pei, Q.B.; Joseph, J. High-speed electrically actuated elastomers with strain greater than 100%. *Science* **2000**, *287*, 836–839. [CrossRef] [PubMed]
6. Zang, J.F.; Ryu, S.; Pugno, N.; Wang, Q.M.; Tu, Q.; Buehler, M.J.; Zhao, X.H. Multifunctionality and control of the crumpling and unfolding of large-area graphene. *Nat. Mater.* **2013**, *12*, 321–325. [CrossRef] [PubMed]
7. Lipomi, D.J.; Vosgueritchian, M.; Tee, B.C.K.; Hellstrom, S.L.; Lee, J.A.; Fox, C.H.; Bao, Z.N. Skin-like pressure and strain sensors based on transparent elastic films of carbon nanotubes. *Nat. Nanotechnol.* **2011**, *6*, 788–792. [CrossRef]
8. Majidi, C.; Kramer, R.; Wood, R.J. A non-differential elastomer curvature sensor for softer-than-skin electronics. *Smart Mater. Struct.* **2011**, *20*, 105017. [CrossRef]
9. Liang, J.J.; Li, L.; Niu, X.F.; Yu, Z.B.; Pei, Q.B. Elastomeric polymer light-emitting devices and displays. *Nat. Photonics* **2013**, *7*, 817–824. [CrossRef]
10. Guo, C.F.; Sun, T.Y.; Liu, Q.H.; Suo, Z.G.; Ren, Z.F. Highly stretchable and transparent nanomesh electrodes made by grain boundary lithography. *Nat. Commun.* **2014**, *5*, 3121. [CrossRef]
11. Wang, J.L.; Wei, J.H.; Su, S.H.; Qiu, J.J.; Wang, S.R. Ion-linked double-network hydrogel with high toughness and stiffness. *J. Mater. Sci.* **2015**, *50*, 5458–5465. [CrossRef]
12. Kim, C.C.; Lee, H.H.; Oh, K.H.; Sun, J.Y. Highly stretchable, transparent ionic touch panel. *Science* **2016**, *353*, 682–687. [CrossRef] [PubMed]
13. Wang, J.L.; Su, S.H.; Qiu, J.J. Biocompatible swelling graphene oxide reinforced double network hydrogels with high toughness and stiffness. *New J. Chem.* **2017**, *41*, 3781–3789. [CrossRef]
14. Wang, J.; Wei, J.; Qiu, J. Facile Synthesis of Tough Double Network Hydrogel. *MRS Adv.* **2016**, *1*, 1953–1958. [CrossRef]
15. Wei, J.; Wang, J.; Su, S.; Wang, S.; Qiu, J. Tough and fully recoverable hydrogels. *J. Mater. Chem. B* **2015**, *3*, 5284–5290. [CrossRef]
16. Wei, J.H.; Su, S.H.; Wang, J.L.; Qiu, J.J. Imitation proteoglycans improve toughness of double network hydrogels. *Mater. Chem. Phys.* **2015**, *166*, 66–72. [CrossRef]
17. Wei, J.H.; Wang, J.L.; Su, S.H.; Wang, S.R.; Qiu, J.J.; Zhang, Z.H.; Christopher, G.; Ning, F.D.; Cong, W.L. 3D printing of an extremely tough hydrogel. *RSC Adv.* **2015**, *5*, 81324–81329. [CrossRef]

18. Markstedt, K.; Mantas, A.; Tournier, I.; Avila, H.M.; Hagg, D.; Gatenholm, P. 3D Bioprinting Human Chondrocytes with Nanocellulose-Alginate Bioink for Cartilage Tissue Engineering Applications. *Biomacromolecules* **2015**, *16*, 1489–1496. [CrossRef]

19. Hull, C.W. Apparatus for Production of Three-Dimensional Objects by Stereolithography. U.S. Patents US4575330A, 11 March 1986.

20. Shusteff, M.; Browar, A.E.M.; Kelly, B.E.; Henriksson, J.; Weisgraber, T.H.; Panas, R.M.; Fang, N.X.; Spadaccini, C.M. One-step volumetric additive manufacturing of complex polymer structures. *Sci. Adv.* **2017**, *3*, 5496. [CrossRef]

21. de Beer, M.P.; van der Laan, H.L.; Cole, M.A.; Whelan, R.J.; Burns, M.A.; Scott, T.F. Rapid, continuous additive manufacturing by volumetric polymerization inhibition patterning. *Sci. Adv.* **2019**, *5*, 8723. [CrossRef]

22. Kelly, B.E.; Bhattacharya, I.; Heidari, H.; Shusteff, M.; Spadaccini, C.M.; Taylor, H.K. Volumetric additive manufacturing via tomographic reconstruction. *Science* **2019**, *363*, 1075–1079. [CrossRef] [PubMed]

23. Murphy, S.V.; Atala, A. 3D bioprinting of tissues and organs. *Nat. Biotechnol.* **2014**, *32*, 773–785. [CrossRef] [PubMed]

24. Lee, V.; Singh, G.; Trasatti, J.P.; Bjornsson, C.; Xu, X.W.; Tran, T.N.; Yoo, S.S.; Dai, G.H.; Karande, P. Design and Fabrication of Human Skin by Three-Dimensional Bioprinting. *Tissue Eng. Part C Methods* **2014**, *20*, 473–484. [CrossRef] [PubMed]

25. Song, S.J.; Choi, J.; Park, Y.D.; Lee, J.J.; Hong, S.Y.; Sun, K. A Three-Dimensional Bioprinting System for Use with a Hydrogel-Based Biomaterial and Printing Parameter Characterization. *Artif. Organs* **2010**, *34*, 1044–1048. [CrossRef]

26. Almeida, C.R.; Serra, T.; Oliveira, M.I.; Planell, J.A.; Barbosa, M.A.; Navarro, M. Impact of 3-D printed PLA- and chitosan-based scaffolds on human monocyte/macrophage responses: Unraveling the effect of 3-D structures on inflammation. *Acta Biomater.* **2014**, *10*, 613–622. [CrossRef]

27. Sun, J.Y.; Zhao, X.H.; Illeperuma, W.R.K.; Chaudhuri, O.; Oh, K.H.; Mooney, D.J.; Vlassak, J.J.; Suo, Z.G. Highly stretchable and tough hydrogels. *Nature* **2012**, *489*, 133–136. [CrossRef]

28. Kamata, H.; Akagi, Y.; Kayasuga-Kariya, Y.; Chung, U.; Sakai, T. "Nonswellable" Hydrogel without Mechanical Hysteresis. *Science* **2014**, *343*, 873–875. [CrossRef]

29. Liu, Z.Y.; Qi, D.P.; Guo, P.Z.; Liu, Y.; Zhu, B.W.; Yang, H.; Liu, Y.Q.; Li, B.; Zhang, C.G.; Yu, J.C.; et al. Thickness-Gradient Films for High Gauge Factor Stretchable Strain Sensors. *Adv. Mater.* **2015**, *27*, 6230–6237. [CrossRef]

30. Kong, H.J.; Lee, K.Y.; Mooney, D.J. Decoupling the dependence of rheological/mechanical properties of hydrogels from solids concentration. *Polymer* **2002**, *43*, 6239–6246. [CrossRef]

31. Chung, J.H.Y.; Naficy, S.; Yue, Z.L.; Kapsa, R.; Quigley, A.; Moulton, S.E.; Wallace, G.G. Bio-ink properties and printability for extrusion printing living cells. *Biomater. Sci.* **2013**, *1*, 763–773. [CrossRef]

32. Lee, J.S.; Hong, J.M.; Jung, J.W.; Shim, J.H.; Oh, J.H.; Cho, D.W. 3D printing of composite tissue with complex shape applied to ear regeneration. *Biofabrication* **2014**, *6*, 024103. [CrossRef] [PubMed]

33. Tabriz, A.G.; Hermida, M.A.; Leslie, N.R.; Shu, W.M. Three-dimensional bioprinting of complex cell laden alginate hydrogel structures. *Biofabrication* **2015**, *7*, 045012. [CrossRef] [PubMed]

34. Mukherjee, D.; Bharath, S. Design of lab model mechanical strength test instrument for tensile strength determination of film formulations. *J. Dent. Orofac. Res.* **2018**, *14*, 18–22.

35. Hong, S.; Sycks, D.; Chan, H.F.; Lin, S.T.; Lopez, G.P.; Guilak, F.; Leong, K.W.; Zhao, X.H. 3D Printing of Highly Stretchable and Tough Hydrogels into Complex, Cellularized Structures. *Adv. Mater.* **2015**, *27*, 4035–4040. [CrossRef] [PubMed]

36. Wang, J.; Lou, L.; Qiu, J. Super-tough hydrogels using ionically crosslinked networks. *J. Appl. Polym. Sci.* **2019**, *136*, 48182. [CrossRef]

37. Chen, Q.; Zhu, L.; Huang, L.N.; Chen, H.; Xu, K.; Tan, Y.; Wang, P.X.; Zheng, J. Fracture of the Physically Cross-Linked First Network in Hybrid Double Network Hydrogels. *Macromolecules* **2014**, *47*, 2140–2148. [CrossRef]

38. Liu, Y.J.; Cao, W.T.; Ma, M.G.; Wan, P.B. Ultrasensitive Wearable Soft Strain Sensors of Conductive, Selfhealing, and Elastic Hydrogels with Synergistic "Soft and Hard" Hybrid Networks. *ACS Appl. Mater. Interfaces* **2017**, *9*, 25559–25570. [CrossRef]

39. Keplinger, C.; Sun, J.Y.; Foo, C.C.; Whitesides, G.M.; Suo, Z.G. Stretchable, Transparent, Ionic Conductors. *Science* **2013**, *341*, 984–987. [CrossRef]

40. Hu, W.; Niu, X.; Li, L.; Yun, S.; Yu, Z.; Pei, Q. Intrinsically stretchable transparent electrodes based on silver-nanowire–crosslinked-polyacrylate composites. *Nanotechnology* **2012**, *23*, 344002. [CrossRef]

41. Yu, Z.B.; Zhang, Q.W.; Li, L.; Chen, Q.; Niu, X.F.; Liu, J.; Pei, Q.B. Highly flexible silver nanowire electrodes for shape-memory polymer lightemitting diodes. *Adv. Mater.* **2011**, *23*, 664–668. [CrossRef]

42. Kim, K.S.; Zhao, Y.; Jang, H.; Lee, S.Y.; Kim, J.M.; Kim, K.S.; Ahn, J.H.; Kim, P.; Choi, J.Y.; Hong, B.H. Large-scale pattern growth of graphene films for stretchable transparent electrodes. *Nature* **2009**, *457*, 706–710. [CrossRef] [PubMed]

43. Hu, L.; Yuan, W.; Brochu, P.; Gruner, G.; Pei, Q. Highly stretchable, conductive, and transparent nanotube thin films. *Appl. Phys. Lett.* **2009**, *94*, 161108. [CrossRef]

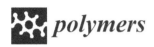

Article

High Electromechanical Deformation Based on Structural Beta-Phase Content and Electrostrictive Properties of Electrospun Poly(vinylidene fluoride-hexafluoropropylene) Nanofibers

Nikruesong Tohluebaji [1], Chatchai Putson [1,2,*] and Nantakan Muensit [1,2]

[1] Department of Physics, Faculty of science, Prince of Songkla University, Songkhla 90110, Thailand; nikruesong.t@pnu.ac.th (N.T.); nantakan.m@psu.ac.th (N.M.)
[2] Center of Excellence in Nanotechnology for Energy (CENE), Songkhla 90110, Thailand
* Correspondence: chatchai.p@psu.ac.th

Received: 23 October 2019; Accepted: 2 November 2019; Published: 5 November 2019

Abstract: The poly(vinylidene fluoride-hexafluoropropylene) (P(VDF-HFP)) polymer based on electrostrictive polymers is essential in smart materials applications such as actuators, transducers, microelectromechanical systems, storage memory devices, energy harvesting, and biomedical sensors. The key factors for increasing the capability of electrostrictive materials are stronger dielectric properties and an increased electroactive β-phase and crystallinity of the material. In this work, the dielectric properties and microstructural β-phase in the P(VDF-HFP) polymer were improved by electrospinning conditions and thermal compression. The P(VDF-HFP) fibers from the single-step electrospinning process had a self-induced orientation and electrical poling which increased both the electroactive β-crystal phase and the spontaneous dipolar orientation simultaneously. Moreover, the P(VDF-HFP) fibers from the combined electrospinning and thermal compression achieved significantly enhanced dielectric properties and microstructural β-phase. Thermal compression clearly induced interfacial polarization by the accumulation of interfacial surface charges among two β-phase regions in the P(VDF-HFP) fibers. The grain boundaries of nanofibers frequently have high interfacial polarization, as they can trap charges migrating in an applied field. This work showed that the combination of electrospinning and thermal compression for electrostrictive P(VDF-HFP) polymers can potentially offer improved electrostriction behavior based on the dielectric permittivity and interfacial surface charge distributions for application in actuator devices, textile sensors, and nanogenerators.

Keywords: electrostrictive properties; actuators; structural β-phase; dielectric properties; P(VDF-HFP) nanofibers; electrospinning; thermal compression

1. Introduction

Electroactive polymers (EAPs) are intelligent materials that convert electrical energy to mechanical energy and vice versa. Common applications of such material include actuators, sensors, energy scavenging, etc. [1]. Electroactive polymers can be classified into two groups which depend on the mechanism responsible for actuation. Electronic EAPs comprise the first group, and the change in range is due to the driven electric field (ferroelectric polymers, dielectric EAPs, electroviscoelastic elastomers, electrostrictive polymers, piezoelectric polymers, etc.) [2]. The second group is composed of ionic EAPs, where the change in shape is due to the mobility or diffusion of ions and their conjugated substances (ionic polymers gels (IPGs), ionic polymer metal composites (IPMCs), conducting polymers, etc.). Electrostrictive polymers are one of the most common electronic EAPs that demonstrate a quadratically based relationship between the strain and electric field. This phenomenon is called electrostriction and occurs in all dielectric materials. It shows a large deformation of electric materials when the electric field is increased. The challenges for electrostrictive performance can be mitigated with a large induced strain under a low electric field. Therefore, improvement of the electrostrictive coefficient is necessary to achieve a high electric field-induced strain. In various studies, it has been suggested that the electrostrictive abilities of a polymer depend on the dielectric permittivity which is a significant parameter. Since the dielectric permittivity directly influences the achievable electrical field-induced strains in actuator applications, an improved electrostrictive polymer needs a high dielectric constant for achieving vast electric field-induced strains. Our previous papers have proposed that the interfacial charge or space charge distribution, which is referred to as a Maxwell–Wagner-type polarization of heterogeneous materials systems, can enhance the electrical and dielectric properties. The electrostriction effect can be observed in the polyurethane (PU) [3] and the family of PVDFs including poly(vinylidene fluoride-trifluoroethylene; P(VDF-TrFE) [4], poly(vinylidene fluoride-trifluoroethylene-chlorofluoro-ethylene); (P(VDF-TrFE-CFE)) [5,6], and poly(vinylidene fluoride-hexafluoropropylene (P(VDF-HFP)) [7].

Poly(vinylidene fluoride-hexafluoropropylene) is a semi-crystalline polymer with the linear formula $(-CH_2CF_2-)_x(-CF_2CF(CF_3)-)_y$ and is a flexible, complex electroactive hydrofluorocarbon polymer that has well-established dielectric properties. Moreover, the features of P(VDF-HFP) are non-toxicity, high stability, its shape and size tailoring ability, and recycling aptitude; these copolymers are gaining momentum in widespread actuator technologies [1]. According to prior literature, Xiaoyan Lu [2] reiterated the strain response in P(VDF HFP) film, and a content of 5% and 15% HFP was measured for electric fields of 0–55 MV/m. Poly(vinylidene fluoride-hexafluoropropylene) can crystallize into α, β, and γ phases [3]. The most common and dominant phase among the three phases, the β-phase, has an orthorhombic structure and an all trans (TTTT) molecule zig-zag conformation. In the β-phase, all the dipoles are aligned in the same direction. It has the most exceptional spontaneous polarization per unit cell which is related to its dielectric properties [4,5]. However, it is challenging to obtain β-P(VDF-HFP). There are more conventional techniques for improving polar β-P(VDF-HFP) which involves electrical poling [6], mechanical extension (drawing) [7], melting processes at high pressure [8], mixing and blending with groups of fillers such as ceramic, clay, or montmorillonite (MMT) [9], and groups of hydrated ionic, magnesium chloride hexahydrate ($MgCl_2·6H_2O$) [10], $Ni(OH)_2$ nanoparticle [11] and groups of conductive nanoparticle, including multiwalled carbon nanotubes (MWNTs) [12,13], carbon nanotubes CNTs [14] and graphene [15]. For example, Swagata Roy et al. [11] presented that the large β-phase of P(VDF-HFP)/MMT up to 85.22% and 82.1% from P(VDF-HFP)/NiMMT composites films. This effect increased the dielectric constant of the P(VDF-HFP)/MMT, and the P(VDF-HFP)/NiMMT film increased due to the increase in MMT content in the matrix of the polymer which exhibited a large interfacial area per unit volume. This is associated with the interfaces of the clay particles and the polymer chains that develop [11]. This result explains that, along with the strong interaction formed between the positive $-CH_2$ dipoles and the negatively charged surface of MMT affection, the vital heightening of the equate that localized the polarization was accompanied with the filler as well as the coupling between the adjoining grains. In addition, the different nanofillers, magnesium chloride hexahydrate ($MgCl_2·6H_2O$), and stretching nanocomposite films also had an effect. The results showed that the P(VDF-HFP)/$MgCl_2·6H_2O$ nanocomposite films, which stretched four-fold, achieved 90% of the β-phase. This experience arises from hydrogen bonding between the ionic interactions and unit chain of the polymer with the hydrated Mg–salt and the polar solvent. Various studies have been undertaken to improve the polar β-phase with the incorporation of conductive elements, for example, MWNTs, CNTs, and graphene [15–21]. The report by Zhou et al. [22] demonstrated the β-phase of P(VDF)/graphene composite nanofibers increased with an increase of graphene 0.1 wt %. The side effects of graphene nanomaterials have improved the stretching effect in the phase transformation of P(VDF) nanofibers. The key factors enhancing the electrostrictive abilities of P(VDF-HFP) are its electroactive β-phase and dielectric permittivity. It was found that P(VDF-HFP) improves the dielectric permittivity because of the intense polarization under an electric field [23]. In this work, we demonstrated the electrostriction of P(VDF-HFP) nanofibers because of its excellent dielectric properties, high surface-area-to-volume ratio, and highly crystallinity. Moreover, the electrostrictive properties of P(VDF-HFP) nanofibers are innovative and worth focusing on. These electrical properties are directly related to permittivity and phase transformation which strongly depend on the surface charge distributions of the material. Increasing the dielectric constant and electroactive β-phase content enhances the electrostrictive coefficient [24]. If the electroactive polymers based on electrostrictive effects include a high dielectric constant, it will likely produce a strong polarization contribution when inducing the external electric field; this generates their large electromechanical deformation. Large electromechanical deformations based on electrostrictive behavior occurred in the high dielectric polymers under their induced polarization contributions when increasing the external electric field strength.

The obtained dipole–dipole interactions in a previous study gave rise to large electrostriction [25]. In this work, we used an electrospinning and thermal compression method to change the geometric morphology of the phase distribution and increase interfacial surface charge distributions.

The selection of the electrospinning condition, including fiber orientation, also supports varying degrees of crystallinity and phase content [26]. It can provide self-induced orientation and electrical poling which increase the electroactive β-crystal phase and dipolar orientation at the same time. In fact, electrospinning can essentially provide a high surface-area-to-volume ratio in electrospun membranes. Several studies have reported that the interfacial charge or space charge distribution, which is referred to as a Maxwell–Wagner–type polarization, can enhance the electrical and dielectric properties in heterogeneous systems [12]. It was found that grain boundaries and interfaces among two regions within a material frequently give rise to interfacial polarization. Thermal compression-induced interfacial polarization occurs owing to the accumulation of interfacial surface charges between two β-phase regions in P(VDF-HFP) fibers.

Therefore, this work set out to study the effects of combining electrospinning and thermal compression of electroactive P(VDF-HFP) nanofibers on their microstructure, crystallinity, β-phase, thermal properties, mechanical properties, electrical and dielectric properties, and, also, their electrostrictive properties with a view to apply them in actuators, textile sensors, nanogenerators, and nanoelectronic devices.

2. Experimental

2.1. Materials

The P(VDF-HFP) powder with 10 wt % HFP (427179, Sigma–Aldrich, Washington, DC, USA) was used as the matrix. The solvent was *N,N*-dimethylformamide (DMF) (D158550, Sigma–Aldrich, Washington, DC, USA). The P(VDF-HFP) solution was prepared as follows. Firstly, 25 wt % of dried P(VDF-HFP) pellets were dissolved in the DMF solvent. The mixture was stirred at 40 °C using a magnetic stirrer in order to achieve a homogeneous solution, as shown in Figure 1a.

2.2. Synthesis of P(VDF-HFP) Films

The P(VDF-HFP) film was fabricated by solution casting (Figure 1b). The P(VDF-HFP) solution was poured onto a glass plate and dried at 80 °C for 3 h. In the end, the P(VDF-HFP) film was formed on the glass substrate.

2.3. Synthesis of P(VDF-HFP) Fiber

The P(VDF-HFP) fiber mats were fabricated by electrospinning (Figure 1c). The viscous P(VDF-HFP) solution was loaded into a 20 mL plastic syringe connected to a stainless-steel needle (gauge 20.5) and fed with a syringe pump (Nz1000 NEWERA Pump Systems Inc., New York, NY, USA) at a flow rate of 1.0 mL h^{-1}. An electric field was generated by a high voltage supply (Trek model 610E, New York, NY, USA), capable of up to 10 kV DC. The positive pole was connected to the steel syringe needle and the collector aluminum plate was the grounded target 11 cm away from the tip of the needle. Porous fibrous films were obtained on the collector plate and were dried overnight at room temperature to evaporate the rest of the solvent. Finally, the fibrous film was compressed with a compression machine (Chareon tut, PR2D-W00L350-PM-WCL-HMI, Samutprakarn, Thailand) at 6.2 MPa and a selected temperature (30, 60 or 80 °C) for 10 min, as shown in Figure 1d. The thickness of all samples was measured using a thickness gauge, and it was in the range of 200 ± 50 μm.

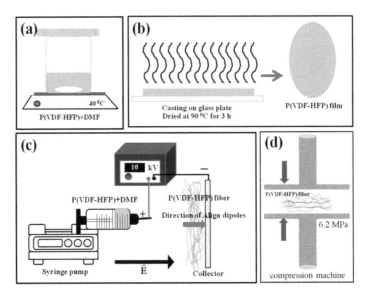

Figure 1. Schematic of the synthesis used with all samples. (**a**) Homogeneous poly(vinylidene fluoride-hexafluoropropylene) (P(VDF-HFP)) solution, (**b**) solution casting, (**c**) electrospinning, and (**d**) fibrous film compression.

2.4. Material Characterization

2.4.1. Surface Topography

The structure and morphology of the P(VDF-HFP) film and fibers were determined by scanning electron microscopy (SEM, FEI Quanta 400, Netherlands). All samples were sputter-coated with gold prior to the SEM imaging. The average fiber diameter and porosity of each sample was analyzed by ImageJ software (National Institutes of Health, 1.46, Madison, WI, USA).

2.4.2. Crystalline Structure and Phase Investigation

The crystalline structure in all samples were examined using an X-ray diffractometer (XRD; X'Pert MPD, Philips, Netherlands) in the 2θ range from 5° to 90° at the scan rate 0.05° s⁻¹ using Cu-K$_\alpha$ radiation (wavelength 0.154 nm) under a voltage of 40 kV. The crystallinity X_c can be estimated as follows [27]:

$$X_c = \frac{\Sigma A_{cr}}{\Sigma A_{cr} + \Sigma A_{amr}} \times 100\% \tag{1}$$

where ΣA_{amr} and ΣA_{cr} are the total integral areas of amorphous halo and crystalline diffraction peaks, respectively. The α and β-phase contents were elucidated from IR spectra obtained with a Fourier transform infrared spectrometer (FTIR-8400S, Shimadzu, Tokyo, Japan). The absorbance data for all samples covered the wavenumber range 400–1000 cm^{-1} with a resolution of 4 cm^{-1}. The fraction of β-phase, F(β) in films or fibers, was calculated using the Lambert–Beer law [10]:

$$F(\beta) = \frac{A_\beta}{\left(\frac{K_\beta}{K_\alpha}\right)A_\alpha + A_\beta} = \frac{A_\beta}{1.26A_\alpha + A_\beta} \tag{2}$$

where A_α and A_β are the absorbance at 764 and 840 cm^{-1}, respectively. $K_\alpha = 6.1 \times 10^4$ cm^2 mol^{-1} and $K_\beta = 7.7 \times 10^4$ cm^2 mol^{-1} are the absorption coefficients at 764 and 840 cm^{-1}, respectively.

The absolute β fraction (%β)is obtained from F(β) and X_c, as in [28]:

$$\%\beta = F(\beta) \times X_c \tag{3}$$

2.4.3. Thermal Analysis

The thermal characteristics melting temperature (T_m), crystallization temperature (T_c), enthalpy of melting (ΔH_m), and enthalpy of crystallization (ΔH_c) were investigated using a differential scanning calorimeter (DSC, Perkin Elmer DSC7, USA). All samples were heated from −100 to 160 °C at a heating rate of 5 °C/min under the N$_2$ atmosphere. Thus, a mass around 10 mg was sealed in an aluminum crucible for DSC-operating experiments.

2.4.4. Mechanical Analysis

The mechanical material properties storage modulus (E'), loss modulus (E) and tan delta (tanδ) were characterized in tensile mode by dynamic mechanical analysis (DMA, Perkin Elmer, Waltham, MA, USA). The elastic behavior is observed from the storage modulus. Tan delta was the ratio of loss modulus to storage modulus. It is often called damping and informs about the energy dissipation in a material. The β-relaxation can be found at −50 °C which corresponds to the segmental motions in the amorphous phase. Also, the α-relaxation is detected above room temperature. The elastic properties of the material and for short-term creep can be recognized from this relaxation [29]. The DMA testing was performed on a temperature ranging from −110 to 160 °C at a heating rate of 5 °C/min with an oscillatory stress amplitude of 0.1 MPa and frequency of 1 Hz to provide the viscoelastic response.

2.4.5. Electrical Properties

The electrical properties, specifically the dielectric constant (ε'_r) which is the real part of the relative permittivity, electrical conductivity (σ), and dielectric losses (tanδ) were evaluated with an LCR meter (IM 3533 HIOKI, Japan) in the frequency range 10^0–10^5 Hz. Each sample, with a thickness of approximately 200 ± 50 μm, was placed among two indium tin oxide electrodes (1 cm diameter) and supplied with 1 V. The dielectric constant can be calculated from [17]:

$$\varepsilon'_r = \frac{Ct}{\varepsilon_0 A} \tag{4}$$

where C, t, A, and ε_0 are the capacitance of the sample, thickenss, the contact area of the electrode, and permittivity of free space (8.854 × 10^{-12} Fm^{-1}), respectively. The electrical conductivity σ is calculated using the equation:

$$\sigma = G\left(\frac{t}{A}\right) = 2\pi f \varepsilon_0 \varepsilon'_r \tan\delta \tag{5}$$

where G is the conductance and f is the applied frequency. The dielectric loss tangent (tanδ) can be estimated from the relationship:

$$\tan\delta = \frac{\varepsilon''_r}{\varepsilon'_r} \tag{6}$$

Here, ε''_r is the imaginary component of relative permittivity.

2.4.6. Electrostrictive Properties

The electrostrictive property of the sample was evaluated by measuring the deformation of strain-induced at low frequency and low electric field strength (f = 1 Hz, E ≤ 3 MV/m) using the photonic displacement apparatus (MTI-2100 Fotonic sensor, New York, USA, sensitivity 5.8 μm/V) setup demonstrated in Figure 2. The dimension of the sample was 3 × 3 cm^2 and the same thickness of 300 μm. The sample was sandwiched among brass electrodes (diameter 2 cm). The electric field (E$_3$) of a high-voltage power supply (Trek model 610E, New York, USA) was applied along the thickness direction of the sample, which was the so-called "3" direction. The electric field-induced strain in polarizable materials was measured in the same direction and, hence, denoted as S$_3$, then the electrostrictive coefficients (M$_{33}$) were given, the relationship can be expressed according to the equation:

$$S_3 = M_{33}E_3^2 \qquad (7)$$

The electrostrictive coefficient can be calculated from the slope of the strain (S$_3$) versus the square of the electric field (E^2$_3$). As a consequence, it can be expressed according to Equation (7).

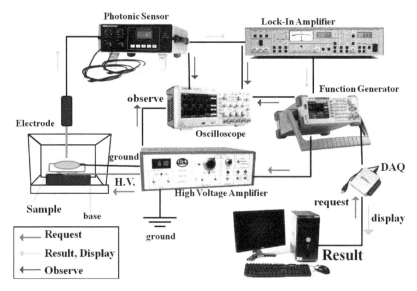

Figure 2. The electrostriction setup.

3. Results and Discussion

3.1. Structure and Morphology

In Figure 3, SEM micrographs display the morphology of P(VDF-HFP) film and fibers. Figure 3a presents the surface of the P(VDF-HFP) film with a 50 μm scale bar showing a clearly smooth, homogeneous, and non-porous surface. The electrospun membranes show a random orientation distribution, high porosity, and smooth and bead-free fibers with an average diameter of 600 ± 50 nm in Figure 3b. Figure 3c–e shows the electrospun P(VDF-HFP) nanofibers after compressing at 30, 60, and 80 °C, respectively. The compression at an elevated temperature reduced porosity and flattened fibers, with an apparent increase in diameter. In Figure 3e, the surface of the fiber mat was almost similar to the film, but with some porosity remaining. The fibers had large contact surfaces which is good for exchanging electric charges. Nanofibers have potential use in the production of sensing devices, because they have a large specific surface that enhances their sensitivity as a sensor [7]. In a previous paper, Kang et al. [30] studied the effect of load compression parameters with electrospun nanofibers at different types of polymers. They compressed poly(caprolactone); PCL and poly(vinyl alcohol); PVA and polyurethane; and PU and nylon nanofibers using a KES-G5s compression tester at different loads (0.5, 1, and 2 N, respectively) under room temperature. They explained that the movement of the fibers was related to the inter-fiber frictions when obtained fibers were passed. The changed density and loss of space between the layers occurred under an applied force. However, the morphology and structure of fibers under compression are not only magnitude force and direction force but also density materials, frictions of fibers, and operating temperature. In our case, the temperature conditions of 30, 60, and 80 °C for P(VDF-HFP) fiber mats were studied under a certain compression force. The morphology of P(VDF-HFP) film and fiber mats depends on the temperature effect, shown in Figure 3, and the changing temperature conditions related to the modification of the interfacial effect within P(VDF-HFP) film and fiber mats. Under the testing conditions, the P(VDF-HFP) fiber mats were strongly melted when the operating temperature increased due to the reduction of the glass temperature.

Figure 3. SEM images of the P(VDF-HFP) (**a**) film, (**b**) fiber, (**c**) 30 °C compressed fibers, (**d**) 60 °C compressed fibers, and (**e**) 80 °C compressed fibers.

3.2. X-ray Diffraction (XRD) Analysis

To confirm the presence of β-phase crystals in P(VDF-HFP) fibers, XRD analysis was performed. Figure 4 displays the XRD patterns generated for the P(VDF-HFP) film, fiber, and fiber mat compressed at 30, 60 or 80 °C. The characteristic reflections of the crystalline phases were seen in the XRD spectrum at $2\theta = 17.9°$ (020) and 26.8° (021) which indicate the large spherulites of the non-polar α-phase crystals while $2\theta = 18.5°$ (110) and 20.1° (110) correspond to the smaller spherulites of γ-phase crystals that co-exist with the α-phase. The specific peaks at $2\theta = 20.3°$ (110) and (200) and 36.7° (020) and (100) correspond to β-phase diffraction [31]. The P(VDF-HFP) film shows the largest non-polar α peak at 26.8° and the smallest β peaks at 20.3° and 36.7°, because it lacks the electrical poling and mechanical stretching treatments. Thus, it shows the α-phase that is commonly the dominant phase in P(VDF-HFP) [32]. After electrospinning, the P(VDF-HFP) fibers and compassed fibers showed a strong β peak (110) at $2\theta = 20.3°$ for the β-phase having an all-trans (TTTT) conformation, while the β peak (020) at $2\theta = 36.7°$ was not clearly observed. Normally, the magnitude of this β peak (020) was quite small when compared with the β peak (110) which may be attributed to the formation of crystalline region and order of polymerization. Therefore, in our work, it was necessary to continue studying by Fourier transform infrared (FT-IR) analysis. It was apparent that the electrospinning successfully formed β-phase crystallites; this should enhance the electrostrictive properties of the nanofibers. The high voltage used during the electrospinning aligned the electric dipoles in the P(VDF-HFP) solution, and the degree of alignment was determined by the applied electric field [23]. In addition, the fibers after pressure and annealing treatments increased the crystalline and β-phases. This demonstrates that the formation of β-phase was induced by electrospinning, pressing, and annealing.

Furthermore, the crystallinity, X_c, of the film and fibers are presented in Table 1. The X_c increased from 49.47% to 55.03% with the compression temperature as shown by the strong peaks for β. The increased crystallinity could be due to the active interactions of the surfaces with polymer chains, inducing the formation of polar β-polymorphs from non-polar α spherulites. In other words, the fraction of crystalline material gradually increased. This appeared to be inferior to the α- to β-phase transformation, since the β-phase strongly depends on the overall crystallinity. However, using only XRD analysis is not enough for comparing the degrees of β-phase transformation, because the α and β peaks are close to each other and the changes are not clear. Therefore, Fourier transform infrared (FT-IR) analysis can provide additional data on the phase structure.

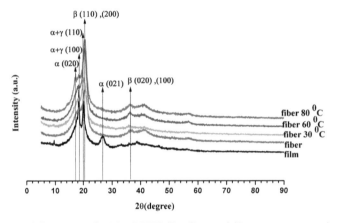

Figure 4. X-ray diffractograms for P(VDF-HFP) film, fiber, and fiber mats compressed at 30°, 60°, and 80 °C.

Table 1. Analysis of the β-phase fraction in the crystalline region of the samples.

Sample	X_c(%)	A_β (cm^{-1})	A_α (cm^{-1})	$F(\beta)$ (%)	%β
Film	49.47	0.3161	0.0875	74.11	36.67
Fiber	49.69	0.1353	0.0176	85.90	42.68
Fiber 30 °C	50.88	0.1569	0.0189	86.80	44.16
Fiber 60 °C	52.44	0.2632	0.0297	87.53	45.90
Fiber 80 °C	55.03	0.327	0.0299	89.65	49.33

3.3. Fourier Transform Infrared Spectroscopy

The FTIR spectra are displayed in Figure 5. These were employed to assess the α- and β-phases, the F(β) of the crystalline region and %β in the samples. According to the literature [32], the non-polar α-phase in P(VDF-HFP) is detected in absorbance bands around 490 cm^{-1} ($-CF_2$ wagging), 530 cm^{-1} ($-CF_2$ bending), 615 cm^{-1} (skeletal bending), 764 cm^{-1} ($-CF_2$ bending), 795 cm^{-1} ($-CH_2$ rocking), and 975 cm^{-1} (twisting) in the IR spectra. In contrast, the large absorbance peaks of the β-phase, attributed to the electroactive polar β-polymorph with a parallel dipole moment, were found at cm^{-1} ($-CF_2$ stretching) and cm^{-1} ($-CH_2$ rocking, $-CF_2$ stretching, and skeletal C−C stretching) in the spectrum.

In the FTIR spectrum, the P(VDF-HFP) film presented the most α-phase at 490 and 764 cm^{-1}. If comparing the IR spectra of the film and nanofiber in Figure 5, it showed a shift of the IR peak of the P(VDF-HFP) film which had a lower wavenumber. The normal α-phase of the P(VDF-HFP) film presented the spectrum peak of −CF$_2$ wagging or out of plane bending and positioned at 490 cm^{-1} [32]. These results are due to the stress and variation in the morphology [33]. Moreover, it may be attributed to a reduction in mass of the molecule polymer chains which depend on the vibration frequency under absorption bands. In the P(VDF-HFP) fiber, all absorbance bands for the α-phase were missing while the absorbance peaks at 509 and 840 cm^{-1} were prominent, signifying a strong emergence of the electroactive β-phase. Therefore, the FTIR results demonstrate that electrospinning promotes the transition to β-phase crystals within the P(VDF-HFP) fibers. In addition, the β-phase of the fiber increased with the compression temperature.

An assessment of the relative fraction of the β-phase content, $F(\beta)$, was executed from the IR spectra using the Lambert–Beer law stated in Equation (2). The $F(\beta)$ for all samples is exhibited in Table 1. The P(VDF-HFP) fiber had F(β) ~85.90% exceeding the film by 11.79%. Electrospinning relies on high electric fields and allows the production of sub-micro to nano-scale fibers, with a β-phase fraction up to 86%without any post-treatment. Furthermore, $F(\beta)$ in the fiber increased from 85.90% to 89.65% when compressed at an elevated temperature. This confirms the positive influence of high pressure on β-phase formation as previously reported. Scheinbeim et al. [14] verified that increasing the quenching pressure from 200 to 700 MPa increased the β-phase content in samples from 0% to 85%.

The absolute β fraction (%β) was estimated from the data of both the X_c and F(β) with Equation (3) as shown in Table 1. About 36.67 %β was obtained in the film, while the largest 49.33% was obtained with a compression at 80 °C of the fiber mat. The emergence of electroactive β-phase was clearly improved by electrospinning and compression at an elevated temperature which was corroborated by FTIR spectra and XRD patterns.

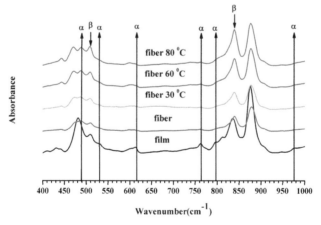

Figure 5. IR spectra of film and fiber P(VDF-HFP) for wavenumbers from 400 to 1000 cm^{-1}.

3.4. Thermal Analysis

The study of the thermal behavior was done using the DSC technique. The data are summarized in Table 2. On comparing the P(VDF-HFP) fiber and film, the onset of the melting (T_m^{on}) and peak melting (T_m^P) temperatures of the fiber increased with the electrospinning. This indicates that the high electric field (and possibly the dimensions of the sample) influenced crystallization in the sample. For compressed the fiber mats, both T_m^{on} and T_m^P decreased with compression temperature. In addition, the final melting temperature (T_m^f) in all cases was in the range from 140 to 180 °C, which corresponds to the melting temperatures of the crystalline phases.

In this analysis, the melting enthalpy (ΔH_m) of compressed P(VDF-HFP) fibers increased with compression temperature, because the particle size increased significantly as seen in the SEM images (Figure 3). Moreover, Madan [34] reported that the specific heat increases as particle size decreases, while the melting entropy and enthalpy decrease. Increased crystallinity can contribute to the mechanical properties of materials. The increased crystallinity may be due to the fact of good interactions and interfacial adhesion between the polymer matrix and the dispersed phase domain surfaces which would also restrict molecular mobility.

The onset crystallization temperature (T_c^{on}), peak crystallization temperature (T_c^p), and final crystallization temperature (T_c^f) decreased with the compression temperature. Elevating the compression temperature reduced the P(VDF-HFP) crystallization temperature progressively, indicating a reduced crystallization rate of the P(VDF-HFP) crystals. Besides, the difference, ΔT_c, increased as the compression temperature decreased. This means that the crystallization rate of the P(VDF-HFP) fibers from the melt was elevated.

Table 2. Thermal properties of the samples.

Sample	T_m^{on}	T_m^p	T_m^f	ΔT_m	ΔH_m	T_c^{on}	T_c^p	T_c^f	ΔT_c	ΔH_c
Film	132.4	158.3	170.5	38.1	21.0	140.5	136.4	131.2	9.3	−26.6
Fiber	136.0	160.1	170.1	34.1	22.7	140.9	137.1	131.7	9.2	−25.2
Fiber 30 °C	143.0	159.0	170.8	27.8	25.2	142.2	136.4	130.8	11.4	−27.0
Fiber 60 °C	137.2	158.3	171.8	34.6	29.3	139.1	134.3	129.1	10.0	−27.0
Fiber 80 °C	135.5	158.2	170.9	35.4	36.3	138.6	134.2	129.6	9.0	−26.2

T_m^{on}: onset melting temperature; T_m^p: peak melting temperature; T_m^f: final melting temperature; $\Delta T_m = T_m^f - T_m^{on}$; ΔH_m: melting enthalpy; T_c^{on}: onset crystallization temperature; T_c^p: peak crystallization temperature; T_c^f: final crystallization temperature; $\Delta T_c = T_c^{on} - T_c^f$; ΔH_c: crystallization enthalpy.

3.5. Mechanical Properties

Dynamic mechanical analysis helps assess the thermomechanical properties and the glass transition temperatures of polymers. The storage modulus (E') and the tan delta as functions of temperature are displayed in Figure 6. The storage modulus decreases with temperature in Figure 6a although not linearly. The storage modulus displays three distinct regions: (1) a glassy high modulus region at low temperatures where the segmental motions are restricted; (2) a transition region with a substantial decrease in E'; (3) and a rubbery area with severe decay in the modulus above the glass transition temperature. The storage modulus in both the glassy and rubbery regions increased due to the thermal compression at 30° to 80 °C and was comparatively high in the glassy region relative to the P(VDF-HFP) fiber. The high storage modulus of 80 °C for the compressed P(VDF-HFP) fiber at low temperatures confirms the reinforcement effect at the molecule interfaces. It can be attributed to the restricted molecular mobility in the P(VDF-HFP) fibers by the strengthened interactions with the polymer matrix [35]. A gradual decrease of E' is observed from −40 to 0 °C which is ascribed to the glass transition of P(VDF-HFP([36]. It can be seen that the compression temperature influenced the glass transition temperature of the P(VDF-HFP) fibers.

Figure 6b presents the loss tangent (tanδ) as a function of temperature for the P(VDF-HFP) fiber and compressed fiber mats. The dielectric relaxation process can be used to explain the ability motions and cross-linking structure in the amorphous phases and crystalline fraction which are related to the dynamic glass transition. Under the relaxation process at lower than room temperature, the β-relaxation for P(VDF-HFP) fiber was −55 °C which can be obtained from the value of the maximum in the loss tangent. It was found that the board of the β-transition for the P(VDF-HFP) fiber presented from −80 to −20 °C. Moreover, it was clearly shown that the β-transition for the P(VDF-HFP) fibers was lower than the fiber mats. This effect occurred in the cooperative segment's mobility of the polymer chains in the amorphous regions. Above the room temperature, the damping relaxation process was observed as two peaks. The first peak damping of the relaxation process provided the α-relaxation which was related to the motions in a crystalline fraction [29], while the second peak relaxation process depicted the melting temperature.

Generally, the glass transition temperature (T_g) of a polymeric material is determined from the peak of the tanδ curve [37]. The tanδ has a peak at approximately −40 °C assigned to the glass transition of pure PVDF. In Figure 6b, the glass transition temperatures (T_g), are approximately −56.17, −44.83, −50.83, and −48.97 for the fiber and the 30, 60 and 80 °C compressed fiber mats, respectively. In fact, the T_g of the polymers was related to the polymer chain's flexibility. When the rigid regions within the polymer increased, it led to an increase in the value of T_g. In the case of fiber mats, the compression process enhanced the rigidity of the polymer based on the crystallinity fraction or hard segments. However, the T_g also depended on the heating or cooling rate and the stress rate.

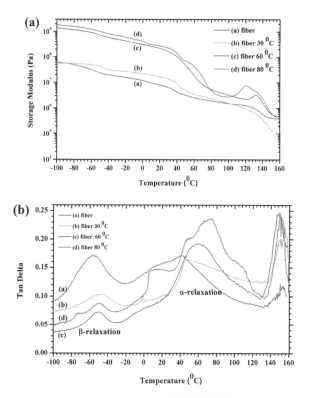

Figure 6. The dynamic mechanical analysis curves of P(VDF-HFP) fiber and compressed fiber mats. (a) Storage modulus and (b) tan delta.

3.6. Electrical Properties

Figure 7a–c presents the dielectric constant (ε_r), loss tangent (tanδ), and conductivity (σ)as functions of frequency from 10^0 to 10^5 Hz for the P(VDF-HFP) film, fibers, and fiber mats compressed at 30°, 60°, and 80 °C. For all samples (Figure 7a), the dielectric constant decreased with frequency. This was because the dipoles of the dielectric materials cannot follow rapid changes in the field direction [38]. At a high frequency, the dielectric constant then only depends on the dipolar polarization, while the alignment of the dipoles lags behind the field in the polymer matrix. The dielectric constant strongly decreased in the low frequency range up to 20 Hz and then suffered a softer decrease at higher frequencies. This was due to the electric polarization inside the matrix which arises from the electrospinning and compression of the P(VDF-HFP) fibers.

All samples had the highest dielectric constant at low frequency, which can be explained by the Maxwell–Wagner polarization in a heterogeneous material [39]. Consequently, the organization of filler within the composites or multilayer dielectric, including electrospun fibers, can enhance the Maxwell–Wagner interfacial polarization with surface charge distribution. The maximum dielectric constant was 8.4 at 1 Hz for fiber mats compressed at 80 °C. Clearly, the thermal compression reduced air gaps and added surface charges causing strong Maxwell–Wagner interfacial polarization.

Interfacial polarization occurs whenever there is an accumulation of charges at interfaces among regions (phases) within a material. Grain boundaries frequently have interfacial polarization, as they can trap charges migrating in an applied field. Dipoles formed by the trapped charges increase the polarization. Interfaces also arise in heterogeneous dielectric materials, for example, when there is a dispersed phase in a continuous matrix. This principle is schematically illustrated in Figure 8. The schematic illustrates the anticipated electroactive β-phase interaction mechanism between the phase and chains of P(VDF-HFP) in the matrix. The P(VDF-HFP) film was prepared without stretching or poling, and the α-P(VDF-HFP) film is shown in Figure 8a. The fibers were then fabricated by electrospinning, thus they contained a β-phase and formed a highly porous fiber mat. Therefore, the interactions among the fiber surfaces were weak because of the air gaps (pores) among the fibers. During thermal compression, the interaction of negatively charged surfaces with C–F and positive –CH$_2$ dipoles from (CH$_2$–CF$_2$) monomers in the P(VDF-HFP) alters the polarity and gives rise to nucleation of the β-phase in fibers [40]. Thus, the compressed fibers had cooperative interactions, apparently with synergistic effects giving an extremely high dielectric constant.

In recent work, the increase in the dielectric constant depended on the crystalline fraction in the polymer [3]. This fraction is normally accompanied by dipole polarization, which increases the melting enthalpy of crystalline domains and can greatly increase the dielectric constant. The observed melting enthalpy and crystallinity were highest for the fiber mats compressed at 80 °C, matching the highest dielectric constant for this case. Moreover, a high specific surface area and overlap without fusing the compressed fibers are the keys to achieving a high dielectric constant [41].

Figure 7b presents the dielectric loss versus frequency. Obviously, the dielectric loss increased with the applied frequency. Large dielectric losses are caused by the charges at high frequencies (10^4 Hz), which is typical owing to the polarization loss and DC conduction loss [42]. On the other hand, the decreased dielectric constant also relates to increased dielectric loss. The AC conductivity in all cases increased at high frequencies, as displayed in Figure 7c. The observed increases in electrical conductivity may be attributed to the polarization of the bound charges [43]. The electrical conductivity linearly increased with frequency, indicating that the number of charge carriers also increased. The electrical conductivity of the fiber mats compressed at 80 °C was the highest, and this might be attributed to the conductive networks formed by the surface contacts in the fibrous matrix and the free electrons within it.

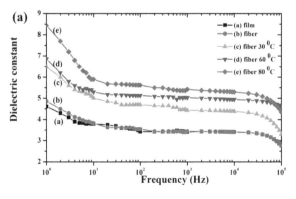

Figure 7. *Cont.*

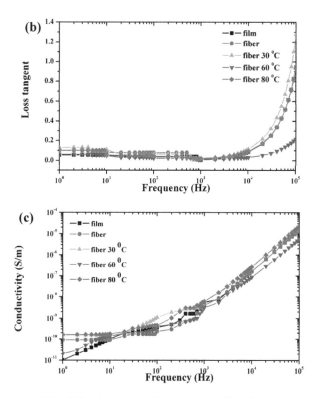

Figure 7. Variation of the dielectric constant (**a**), loss tangent (**b**) and AC conductivity (**c**) with frequencies from 10^0 to 10^5 Hz for the film, fibers, and compressed fiber mats.

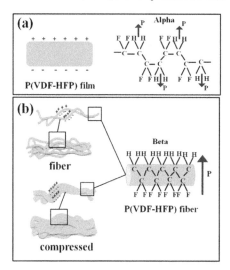

Figure 8. Schematic of the proposed β-phase transformation mechanism.

3.7. Electrostrictive Properties

The longitudinal strain (S_3) behavior of the P(VDF-HFP) film, fibers, and fiber mats compressed at 30°, 60°, and 80°C and induced by an external electric field (E_3) at a frequency of 1 Hz, is presented in Figure 9a. At a low electric field, the electrostriction behavior demonstrated that the total thickness strain had an approximately quadratic relation to the applied electric field. This can be illustrated by the electrostriction which is shown in Equation (7) and Maxwell stress. The Maxwell stress effect involves electrostatic attractions and interactions with the charges on the electrodes showing $S_M = \varepsilon_0 \varepsilon_r E_3^2 / Y$. This equation can be used to estimate the value of the Maxwell stress which can be neglected due to its small and negligible value in the case of a low dielectric constant at an applied low electric field. The induced strain of all samples based on the electrostrictive behavior can be observed. The linear relationships between the induced strain and the electric field, shown in Figure 9b, display that their slope can be assigned as the electrostrictive coefficient (M_{33}). As a result, the electrostrictive behavior of the samples can be expressed according to Equation (7). The M_{33} increased with the fraction of temperatures compressed and the increased number of polymer–polymer interfaces may play a key role, affecting both the electrical and mechanical properties as shown in Figure 9c. This was reflected by the abruptly increased space charge distribution, and the increasing dielectric constant increased with the compressed fibers. Clearly, this study indicates that fibers compressed by temperature are promising electrostrictive materials for actuation and can be fabricated with a simple preparation.

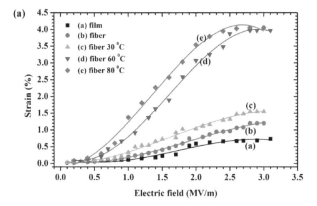

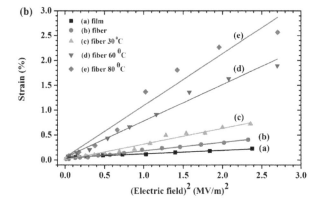

Figure 9. *Cont.*

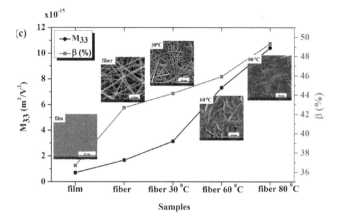

Figure 9. Strain behaviors of film, fibers, and compressed fibers as a function of the (**a**) electric field and (**b**) square of the electric field at 1 Hz. (**c**) The effect of the compression fibers on the electrostrictive coefficients and absolute β fraction.

4. Conclusions

In conclusion, electrospinning fibers and thermal compression of the fiber mats enhanced the availability of the interfacial charges, dielectric constant, electroactive β-phase, and electrostrictive coefficient content in P(VDF-HFP). The high electrostatic field in the electrospinning process caused orientation polarization, which apparently helped transform non-polar α-phase to electroactive β-phase in the formed fibers. Increasing the voltage during electrospinning increased the β-phase fraction in fibers from 74.11% to 85.90%. In this study, compressing P(VDF-HFP) fiber mats at 80 °C gave the highest, 89.65%, β-phase fraction among the cases tested. In addition, the dielectric constant and the crystallinity increased with the compression temperature up to 80 °C. This case gave the maximal observed dielectric constant of 8.4 at 1 Hz and also had the largest absolute β fraction (%β) of about 49.33% among the cases tested. Thus, the electrospinning and thermal compressing coupling processes can enhance the induced interfacial polarization, and β-phase leads to the high electrostrictive properties of obtained P(VDF-HFP) nanofibers.

Author Contributions: N.T. prepared and fabricated all samples. N.T. and C.P. planned and performed the experiments, characterization and data analysis. N.T., C.P., and N.M. wrote the text and reviewed the manuscript. N.M. and C.P. supported equipment and supervision.

Funding: This research was supported by the Science Achievement Scholarship of Thailand and research funding from graduate school, Prince of Songkla University, Thailand.

Acknowledgments: The authors are profoundly grateful to the Department of Physics and the Center of Excellence in Nanotechnology for Energy at the Prince of Songkla University for equipment and other support.

Conflicts of Interest: The authors declare no conflict of interest.

References

1. Brochu, P.; Pei, Q. Advances in dielectric elastomers for actuators and artificial muscles. *Macromol. Rapid Commun.* **2010**, *31*, 10–36. [CrossRef] [PubMed]
2. Lang, S.B.; Muensit, S. Review of some lesser-known applications of piezoelectric and pyroelectric polymers. *Appl. Phys. A* **2006**, *85*, 125–134. [CrossRef]
3. Putson, C.; Muensit, N. High electromechanical performance of modified electrostrictive polyurethane three-phase composites. *Compos. Sci. Technol.* **2018**, *158*, 164–174.

4.   Bharti, V.; Cheng, Z.Y.; Gross, S.; Xu, T.B.; Zhang, Q.M. High electrostrictive strain under high mechanical stress in electron-irradiated poly(vinylidene fluoride-trifluoroethylene) copolymer. *Appl. Phys. Lett.* **1999**, *75*, 2653–2655. [CrossRef]

5.   Cottinet, P.J.; Lallart, M.; Guyomar, D.; Guiffard, B.; Lebrun, L.; Sebald, G.; Putson, C. Analysis of ac-dc conversion for energy harvesting using an electrostrictive polymer P(VDF-TrFE-CFE). *IEEE Trans. Ultrason. Ferroelectr. Freq. Control* **2011**, *58*, 30–42. [CrossRef] [PubMed]

6.   Xia, F.; Cheng, Z.Y.; Xu, H.S.; Li, H.F.; Zhang, Q.M.; Kavarnos, G.J.; Ting, R.Y.; Abdul-Sadek, G.; Belfield, K.D. High Electromechanical Responses in a Poly(vinylidene fluoride-trifluoroethylene-chlorofluoroethylene) Terpolymer. *Adv. Mater.* **2002**, *14*, 1574–1577. [CrossRef]

7.   Lu, X.; Schirokauer, A.; Scheinbeim, J. Giant electrostrictive response in poly(vinylidene fluoride-hexafluoropropylene) copolymers.). *IEEE Trans. Ultrason. Ferroelectr. Freq. Control* **2000**, *47*, 1291–1295. [CrossRef]

8.   Wu, L.; Yuan, W.; Hu, N.; Wang, Z.; Chen, C.; Qiu, J.; Ying, J.; Li, Y. Improved piezoelectricity of PVDF-HFP/carbon black composite films. *J. Phys. D Appl. Phys.* **2014**, *47*, 135302. [CrossRef]

9.   Martins, P.; Lopes, A.C.; Lanceros-Mendez, S. Electroactive phases of poly(vinylidene fluoride): Determination, processing and applications. *Prog. Polym. Sci.* **2014**, *39*, 683–706. [CrossRef]

10. Sencadas, V.; Gregorio, R.; Lanceros-Méndez, S. α to β Phase Transformation and Microestructural Changes of PVDF Films Induced by Uniaxial Stretch. *J. Macromol. Sci. Part B* **2009**, *48*, 514–525. [CrossRef]

11. Thakur, P.; Kool, A.; Bagchi, B.; Hoque, N.A.; Das, S.; Nandy, P. Improvement of electroactive β phase nucleation and dielectric properties of WO3·H2O nanoparticle loaded poly(vinylidene fluoride) thin films. *RSC Adv.* **2015**, *5*, 62819–62827. [CrossRef]

12. Sukwisute, P.; Muensit, N.; Soontaranon, S.; Rugmai, S. Micropower energy harvesting using poly(vinylidene fluoride hexafluoropropylene). *Appl. Phys. Lett.* **2013**, *103*, 063905. [CrossRef]

13. Wang, F.; Frubing, P.; Wirges, W.; Gerhard, R.; Wegener, M. Enhanced Polarization in Melt-quenched and Stretched Poly(vinylidene Fluoride-Hexafluoropropylene) Films. *IEEE Trans. Dielectr. Electr. Insul.* **2010**, *17*, 1088–1095. [CrossRef]

14. Scheinbeim, J.; Nakafuku, C.; Newman, B.A.; Pae, K.D. High-pressure crystallization of poly(vinylidene fluoride). *J. Appl. Phys.* **1979**, *50*, 4399–4405. [CrossRef]

15. Ma, Y.; Tong, W.; Wang, W.; An, Q.; Zhang, Y. Montmorillonite/PVDF-HFP-based energy conversion and storage films with enhanced piezoelectric and dielectric properties. *Compos. Sci. Technol.* **2018**, *168*, 397–403. [CrossRef]

16. Yuennan, J.; Sukwisute, P.; Boripet, B.; Muensit, N. Phase Transformation, Surface Morphology and Dielectric Property of P(VDF-HFP)/MgCl2·6H2O Nanocomposites. *J. Phys. Conf. Ser.* **2017**, *901*, 012085. [CrossRef]

17. Roy, S.; Thakur, P.; Hoque, N.A.; Bagchi, B.; Das, S. Enhanced electroactive β-phase nucleation and dielectric properties of PVdF-HFP thin films influenced by montmorillonite and Ni(OH)2 nanoparticle modified montmorillonite. *RSC Adv.* **2016**, *6*, 21881–21894. [CrossRef]

18. Zhou, T.; Zha, J.W.; Hou, Y.; Wang, D.; Zhao, J.; Dang, Z.M. Surface-functionalized MWNTs with emeraldine base: Preparation and improving dielectric properties of polymer nanocomposites. *ACS Appl. Mater. Interfaces* **2011**, *3*, 4557–4560. [CrossRef]

19. Wongtimnoi, K.; Guiffard, B.; Bogner-Van de Moortele, A.; Seveyrat, L.; Cavaillé, J. Electrostrictive thermoplastic polyurethane-based nanocomposites filled with carboxyl-functionalized multi-walled carbon nanotubes (MWCNT-COOH): Properties and improvement of electromechanical activity. *Compos. Sci. Technol.* **2013**, *85*, 23–28. [CrossRef]

20. Yan, J.; Jeong, Y.G. Roles of carbon nanotube and BaTiO 3 nanofiber in the electrical, dielectric and piezoelectric properties of flexible nanocomposite generators. *Compos. Sci. Technol.* **2017**, *144*, 1–10. [CrossRef]

21. Tarhini, A.A.; Tehrani-Bagha, A.R. Graphene-based polymer composite films with enhanced mechanical properties and ultra-high in-plane thermal conductivity. *Compos. Sci. Technol.* **2019**, *184*, 107797. [CrossRef]

22. Abolhasani, M.M.; Shirvanimoghaddam, K.; Naebe, M. PVDF/graphene composite nanofibers with enhanced piezoelectric performance for development of robust nanogenerators. *Compos. Sci. Technol.* **2017**, *138*, 49–56. [CrossRef]

23. Dhakras, D.; Borkar, V.; Ogale, S.; Jog, J. Enhanced piezoresponse of electrospun PVDF mats with a touch of nickel chloride hexahydrate salt. *Nanoscale* **2012**, *4*, 752–756. [CrossRef] [PubMed]

24. Sharma, M.; Srinivas, V.; Madras, G.; Bose, S. Outstanding dielectric constant and piezoelectric coefficient in electrospun nanofiber mats of PVDF containing silver decorated multiwall carbon nanotubes: Assessing through piezoresponse force microscopy. *RSC Adv.* **2016**, *6*, 6251–6258. [CrossRef]

25. Wongtimnoi, K.; Guiffard, B.; Bogner-Van de Moortele, A.; Seveyrat, L.; Gauthier, C.; Cavaillé, J.Y. Improvement of electrostrictive properties of a polyether-based polyurethane elastomer filled with conductive carbon black. *Compos. Sci. Technol.* **2011**, *71*, 885–892. [CrossRef]

26. Ribeiro, C.; Sencadas, V.; Ribelles, J.L.G.; Lanceros-Méndez, S. Influence of Processing Conditions on Polymorphism and Nanofiber Morphology of Electroactive Poly(vinylidene fluoride) Electrospun Membranes. *Soft Mater.* **2010**, *8*, 274–287. [CrossRef]

27. Karan, S.K.; Mandal, D.; Khatua, B.B. Self-powered flexible Fe-doped RGO/PVDF nanocomposite: An excellent material for a piezoelectric energy harvester. *Nanoscale* **2015**, *7*, 10655–10666. [CrossRef]

28. Low, Y.K.A.; Tan, L.Y.; Tan, L.P.; Boey, F.Y.C.; Ng, K.W. Increasing solvent polarity and addition of salts promote β-phase poly(vinylidene fluoride) formation. *J. Appl. Polym. Sci.* **2013**, *128*, 2902–2910. [CrossRef]

29. Mano, J.F.; Sencadas, V.; Costa, A.M.; Lanceros-Méndez, S. Dynamic mechanical analysis and creep behaviour of β-PVDF films. *Mater. Sci. Eng. A* **2004**, *370*, 336–340. [CrossRef]

30. Kang, J.; Sukigara, S. Development of bulk compression measurement for nonwoven sheets of electrospun nanofibers. *Textile Res. J.* **2013**, *83*, 1524–1531. [CrossRef]

31. Patro, T.U.; Mhalgi, M.V.; Khakhar, D.V.; Misra, A. Studies on poly(vinylidene fluoride)–clay nanocomposites: Effect of different clay modifiers. *Polymer* **2008**, *49*, 3486–3499. [CrossRef]

32. Prabakaran, K.; Mohanty, S.; Nayak, S.K. Influence of surface modified nanoclay on electrochemical properties of PVDF-HFP composite electrolytes. *Int. J. Plast. Technol.* **2014**, *18*, 349–361. [CrossRef]

33. Du, C.-H.; Zhu, B.-K.; Xu, Y.-Y. The effects of quenching on the phase structure of vinylidene fluoride segments in PVDF-HFP copolymer and PVDF-HFP/PMMA blends. *J. Mater. Sci.* **2006**, *41*, 417–421. [CrossRef]

34. Babu, K.S.; Reddy, A.R.; Sujatha, C.; Reddy, K.V.; Mallika, A.N. Synthesis and optical characterization of porous ZnO. *J. Adv. Ceram.* **2013**, *2*, 260–265. [CrossRef]

35. Singh, M.; Lara, S.O.; Tlali, S. Effects of size and shape on the specific heat, melting entropy and enthalpy of nanomaterials. *J. Taibah Univ. Sci.* **2018**, *11*, 922–929. [CrossRef]

36. Bikiaris, D.N.; Nianias, N.P.; Karagiannidou, E.G.; Docoslis, A. Effect of different nanoparticles on the properties and enzymatic hydrolysis mechanism of aliphatic polyesters. *Polym. Degrad. Stab.* **2012**, *97*, 2077–2089. [CrossRef]

37. George, K.E.; Komalan, C.; Kumar, P.A.S.; Varughese, K.T.; Thomas, S. Dynamic mechanical analysis of binary and ternary polymer blends based on nylon copolymer/EPDM rubber and EPM grafted maleic anhydride compatibilizer. *Express Polym. Lett.* **2007**, *1*, 641–653.

38. Prabakaran, K.; Mohanty, S.; Nayak, S.K. Improved electrochemical and photovoltaic performance of dye sensitized solar cells based on PEO/PVDF–HFP/silane modified TiO2 electrolytes and MWCNT/Nafion®counter electrode. *RSC Adv.* **2015**, *5*, 40491–40504. [CrossRef]

39. Mahapatra, S.S.; Yadav, S.K.; Yoo, H.J.; Cho, J.W. Highly stretchable, transparent and scalable elastomers with tunable dielectric permittivity. *J. Mater. Chem.* **2011**, *21*, 7686. [CrossRef]

40. Fan, B.H.; Zha, J.W.; Wang, D.R.; Zhao, J.; Dang, Z.M. Experimental study and theoretical prediction of dielectric permittivity in BaTiO3/polyimide nanocomposite films. *Appl. Phys. Lett.* **2012**, *100*, 092903. [CrossRef]

41. Xing, C.; Guan, J.; Li, Y.; Li, J. Effect of a room-temperature ionic liquid on the structure and properties of electrospun poly(vinylidene fluoride) nanofibers. *ACS Appl. Mater. Interfaces* **2014**, *6*, 4447–4457. [CrossRef] [PubMed]

42. Jana, S.; Garain, S.; Sen, S.; Mandal, D. The influence of hydrogen bonding on the dielectric constant and the piezoelectric energy harvesting performance of hydrated metal salt mediated PVDF films. *Phys. Chem. Chem. Phys.* **2015**, *17*, 17429–17436. [CrossRef] [PubMed]

43. Wang, Z.; Wang, T.; Fang, M.; Wang, C.; Xiao, Y.; Pu, Y. Enhancement of dielectric and electrical properties in BFN/Ni/PVDF three-phase composites. *Composites Sci. Technol.* **2017**, *146*, 139–146. [CrossRef]

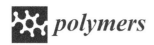

Article

A Multi-Parameter Perturbation Solution and Experimental Verification for Bending Problem of Piezoelectric Cantilever Beams

Zhi-Xin Yang [1], Xiao-Ting He [1,2,*], Hong-Xia Jing [1] and Jun-Yi Sun [1,2]

[1] School of Civil Engineering, Chongqing University, Chongqing 400045, China;
20141602063@cqu.edu.cn (Z.-X.Y.); jinghongxiajy@163.com (H.-X.J.); sunjunyi@cqu.edu.cn (J.-Y.S.)
[2] Key Laboratory of New Technology for Construction of Cities in Mountain Area, Chongqing University,
Ministry of Education, Chongqing 400045, China
* Correspondence: hexiaoting@cqu.edu.cn; Tel.: +86-(0)23-65120720

Received: 31 October 2019; Accepted: 21 November 2019; Published: 24 November 2019

Abstract: The existing studies indicate that the application of piezoelectric polymers is becoming more and more extensive, especially in the analysis and design of sensors or actuators, but the problems of piezoelectric structure are usually difficult to solve analytically due to the force–electric coupling characteristics. In this study, the bending problem of a piezoelectric cantilever beam was investigated via theoretical and experimental methods. First, the governing equations of the problem were established and non-dimensionalized. Three piezoelectric parameters were selected as perturbation parameters and the perturbation solution of the equations was finally obtained using a multi-parameter perturbation method. In addition, the relevant experiments of the piezoelectric cantilever beam were carried out, and the experimental results were in good agreement with the theoretical solutions. Based on the experimental results, the effect of piezoelectric properties on the bending deformation of piezoelectric cantilever beams was analyzed and discussed. The results indicated that the multi-parameter perturbation solution obtained in this study is effective and it may serve as a theoretical reference for the design of sensors or actuators made of piezoelectric polymers.

Keywords: multi-parameter perturbation method; piezoelectric polymers; experimental verification; cantilever beam; force–electric coupling characteristics

---

1. Introduction

Piezoelectric polymers have been widely used in sensors, actuators, electronic information and intelligent structures because of its great force–electric coupling characteristics [1–6]. The piezoelectric polymers usually participate in the work of piezoelectric instruments in the form of piezoelectric sheets which usually are simplified to a piezoelectric cantilever beam [7–9]. The problems of piezoelectric cantilever beams are usually difficult to be solved analytically due to the existence of the force–electric coupling constitutive relation. It is known that the design of piezoelectric instruments often requires the analytical expression of the problem of piezoelectric cantilever beams as a theoretical reference. Therefore, it is necessary and meaningful to find an efficient analytical method for solving the problem of piezoelectric cantilever beams and giving their analytical solutions.

In the past twenty years, many researchers have studied the problem of piezoelectric cantilever beams and obtained some corresponding solutions. Wang and Chen [10] obtained a general solution of the control equation for the three-dimensional problem of transverse isotropic piezoelectric material by means of a set of new potential functions representing displacement component and potential function, and solved the problem of spatial piezoelectric material under the action of concentrated transverse shear force. Lin et al. [11] derived the analytical expressions of displacement, potential, and stress distribution of piezoelectric beams which were simply supported at both ends under a uniform load. According to the plane stress problem, Mei and Zeng [12] directly derived the equation of state of piezoelectric beams from the piezoelectric physical equation, and on this basis, the exact state equation solution of electromechanical coupling effect of simply supported piezoelectric beams at both ends under a uniform load was given. On the basis of three-dimensional constitutive equations and their simplified equations of elastic piezoelectric materials, Zhu [13] derived the analytic solution to a piezoelectric cantilever beam with concentrated force at the free end in terms of displacements and voltage. For the orthotropic piezoelectric plane problem, Ding et al. [14–16] solved a series of piezoelectric beam problems and obtained the corresponding exact solutions with the trial and error method on the basis of the general solution in the case of three distinct eigenvalues, and expressed all displacements, electrical potential, stresses, and electrical displacements by three displacement functions in terms of harmonic polynomials. Yang and Liu [17] investigated the bending of transversely isotropic cantilever beams under an end load, and derived the simplified linear elastic equations of piezoelectric cantilever beams according to the characters of the problem. Pang et al. [18] manufactured a typical Li- and Ta/Sb-modified, alkaline niobate-based, lead-free piezoelectric ceramics by two-step sintering and investigated the sintering condition dependence of dielectric constants and piezoelectric properties. Zhu et al. [19] studied the active vibration control of piezoelectric cantilever beams, where an adaptive feed forward controller (AFC) was utilized to reject the vibration with unknown multiple frequencies. Peng et al. [20] presented time-delayed feedback control to reduce the non-linear resonant vibration of a piezoelectric elastic beam and examined three single-input linear time-delayed feedback control methodologies: displacement, velocity, and acceleration time-delayed feedback. Liu and Yang [21] studied the bending problem of a cantilever beam made of a transversely isotropic piezoelectricity medium under uniformly distributed loads. Shi et al. [22,23] studied the analytical solution of a density functionally gradient piezoelectric cantilever under axial and transverse uniform loads and applied DC voltages and then, solved the force–electric coupling plane strain problem of simply supported beams under a uniform load by the inverse method. Wang et al. [24] dealt with the vibration analysis of a circular plate surface bonded by two piezoelectric layers, based on the Kirchhoff plate model. Recently, Lian et al. [25] studied the problem of a functionally graded piezoelectric cantilever beam under combined loads, but non-dimensionalization was not considered in solving the problem. There is still a lot of research performed in this field, which will not be elaborated here. The summation of results of existing research shows that there are still some unsolved problems. First, non-dimensionalization was not considered in the existing research. We know that piezoelectric materials have not only mechanical properties, but also electrical properties. So, there are both mechanical units and electrical units to be solved, which may lead to computational errors. Second, the existing research basically provides theoretical solutions, but there are a few related experimental verifications. Therefore, the reliability of theoretical solutions cannot be guaranteed. Besides, there has been no unified and effective method for solving the problems of piezoelectric structure.

Parameter perturbation method is a general analysis method for solving approximate solutions of non-linear mechanical problems. It has been successfully applied to various fields of non-linear structural analysis, such as non-linear bending and post-buckling, and has become a powerful tool for solving non-linear problems of structures. Generally speaking, the perturbation method is based on a selected small parameter. In order to solve the problem of parameter selection, Chen and Li [26] put forward the concept of free parameter perturbation method, that is, there is no need to point out the physical meaning of perturbation parameters during perturbation, which provides a new idea for solving the parameter selection problem of parameter perturbation method. Lian et al. [27] solved the Hencky membrane problem without a small-rotation-angle assumption by the single-parameter perturbation method. The successful application of perturbation method depends, to a large extent, on the reasonable choice of small parameters, but the selection of perturbation parameters does not have a set of step-by-step procedures, which can only rely on deep understanding and multiple attempts. To avoid the difficulty in the selection of perturbation parameters, researchers can select multiple parameters, that is, the so-called "multi-parameter perturbation method". For multi-parameter perturbation method, Nowinski and Ismail [28] solved the cylindrical orthotropic circular plate problem under a uniform load by using the two-parameter perturbation method. The application of the multi-parameter perturbation method in beam problem was proposed by Chien [29] in 2002, the classical Euler–Bernoulli equation of bending beams was solved by using load and beam height differences as perturbation parameters. Later, He and Chen [30] simplified the bending moment by using the quasi-linear analysis method, so that the parameter perturbation process was directly aimed at the algebra equation rather than the integral equation, and the two-parameter perturbation solution of the large deflection bending problem of a cantilever beam was obtained, and the integrity of the two-parameter perturbation solution was analyzed. Recently, He et al. [31,32] comprehensively analyzed the large deflection problem of beams with height difference under various boundary conditions, put forward the so-called "two-parameter perturbation method", and successfully applied this method to the solution of bimodular von-Kármán thin plate equation. But so far, the perturbation method of three or more parameters has only a few reports.

In this study, we will derive the theoretical solution of the bending problem of piezoelectric cantilever beams by the multi-parameter perturbation method. The whole paper is organized as follows. In Section 2, the mechanical model of the problem solved here will be established, and the governing equations will be given and dimensionless. In Section 3, the three piezoelectric parameters will be selected as perturbation parameters, and the dimensionless governing equations will be solved by the multi-parameter perturbation method. The solution presented in this paper will be compared with the existing analytical solution from Yang and Liu [17] in Section 4. Next, in Section 5, we will show the related experiments of the piezoelectric cantilever beam, compare the experimental results with the solution presented here, and also discuss the effect of the piezoelectric properties on the deformation of piezoelectric cantilever beams. According to the results mentioned above, some main conclusions will be drawn in Section 6.

2. Mechanical Model and Basic Equations

In this study, the mechanical model of the transversely isotropic piezoelectric cantilever beam is established by using two-dimensional elastic beam theory and neglecting shear deformation. As shown in Figure 1, an transversely isotropic piezoelectric cantilever beam is fixed at its right end and subjected to a uniformly distributed load q on its upper surface, a concentrated force P and a bending moment M at its left end, in which l, b, and h denote the length, width, and height of the beam, respectively, and O denotes the origin of the coordinates. A rectangular coordinate system is introduced with the upper and lower surfaces of the beam lying in $z = -h/2$ and $z = h/2$.

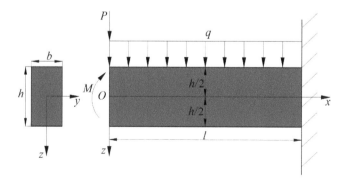

Figure 1. Scheme of a piezoelectric cantilever beam.

Supposing that the polarization direction is the forward direction of the z-axis, let us take a microelement in the piezoelectric cantilever beam, and from the balance of the force, we may obtain, by neglecting the body force

$$
\begin{cases}
\frac{\partial \sigma_x}{\partial x} + \frac{\partial \tau_{zx}}{\partial z} = 0 \\
\frac{\partial \tau_{zx}}{\partial x} + \frac{\partial \sigma_z}{\partial z} = 0
\end{cases},
\tag{1}
$$

where σ_x, σ_z and τ_{zx} are the stress components. The equation of Maxwell electric displacement conservation is

$$
\frac{\partial D_x}{\partial x} + \frac{\partial D_z}{\partial z} = 0,
\tag{2}
$$

where D_x and D_z are the electric displacement components. The constitutive equations of piezoelectric polymeric materials considered are

$$
\begin{cases}
\varepsilon_x = s_{11}\sigma_x + s_{13}\sigma_z + d_{31}E_z \\
\varepsilon_z = s_{13}\sigma_x + s_{33}\sigma_z + d_{33}E_z \\
\gamma_{zx} = s_{44}\tau_{zx} + d_{15}E_x \\
D_x = d_{15}\tau_{zx} + \lambda_{11}E_x \\
D_z = d_{31}\sigma_x + d_{33}\sigma_z + \lambda_{33}E_z
\end{cases},
\tag{3}
$$

where ε_x, ε_z, and γ_{zx} are the strain components; and E_x and E_z are the electric field intensity components. The geometric equations of the piezoelectric cantilever beam are

$$
\varepsilon_x = \frac{\partial u}{\partial x}, \varepsilon_z = \frac{\partial w}{\partial z}, \gamma_{zx} = \frac{\partial u}{\partial z} + \frac{\partial w}{\partial x},
\tag{4}
$$

where u and w are the displacement components. From Equation (4), the strain consistency equation is obtained as follows:

$$
\frac{\partial^2 \varepsilon_x}{\partial z^2} + \frac{\partial^2 \varepsilon_z}{\partial x^2} - \frac{\partial^2 \gamma_{zx}}{\partial z \partial x} = 0.
\tag{5}
$$

The relationship between electric field intensity and electric potential are

$$
E_x = -\frac{\partial \phi}{\partial x}, E_z = -\frac{\partial \phi}{\partial z},
\tag{6}
$$

where ϕ is the electric potential function. By introducing the Airy stress function $U(x,z)$, the stress components can be expressed as

$$
\sigma_x = \frac{\partial^2 U}{\partial z^2}, \sigma_z = \frac{\partial^2 U}{\partial x^2}, \tau_{zx} = -\frac{\partial^2 U}{\partial z \partial x}.
\tag{7}
$$

The boundary conditions of the problem of the piezoelectric cantilever beam are

$$
\begin{cases}
\int_{-h/2}^{h/2} \tau_{zx}dz = \int_{-h/2}^{h/2} \frac{\partial^2 U}{\partial z \partial x}dz = -\frac{P}{b}, \\
\int_{-h/2}^{h/2} \sigma_x dz = \int_{-h/2}^{h/2} \frac{\partial^2 U}{\partial z^2}dz = 0, \\
\int_{-h/2}^{h/2} z\sigma_x dz = \int_{-h/2}^{h/2} z\frac{\partial^2 U}{\partial z^2}dz = \frac{M}{b}
\end{cases}, \text{ at } x = 0,
\tag{8}
$$

$$
\sigma_z = \frac{\partial^2 U}{\partial x^2} = 0, \tau_{zx} = -\frac{\partial^2 U}{\partial x \partial z} = 0, \text{ at } z = h/2,
\tag{9}
$$

$$
\sigma_z = \frac{\partial^2 U}{\partial x^2} = -q, \tau_{zx} = -\frac{\partial^2 U}{\partial x \partial z} = 0, \text{ at } z = -h/2,
\tag{10}
$$

$$
\int_{-h/2}^{h/2} D_x dz = 0, \text{ at } x = 0 \text{ and } x = l,
\tag{11}
$$

$$E_z = \frac{\partial \phi}{\partial z} = 0, \text{ at } z = \pm h/2, \tag{12}$$

and

$$u = 0, w = 0, \frac{\partial w}{\partial x} = 0, \text{ at } z = 0 \text{ and } x = l. \tag{13}$$

Substituting Equations (3), (6), and (7) into Equations (2) and (4), we may obtain two equations of the stress function $U(x, z)$ and the potential function ϕ

$$d_{31}\frac{\partial^3 U}{\partial z^3} + d_{33}\frac{\partial^3 U}{\partial x^2 \partial z} - d_{15}\frac{\partial^3 U}{\partial x^2 \partial z} = \lambda_{33}\frac{\partial^2 \phi}{\partial z^2} + \lambda_{11}\frac{\partial^2 \phi}{\partial x^2} \tag{14}$$

and

$$s_{11}\frac{\partial^4 U}{\partial z^4} + (2s_{13} + s_{44})\frac{\partial^4 U}{\partial x^2 \partial z^2} + s_{33}\frac{\partial^4 U}{\partial x^4} = d_{31}\frac{\partial^3 \phi}{\partial z^3} + d_{33}\frac{\partial^3 \phi}{\partial x^2 \partial z} - d_{15}\frac{\partial^3 \phi}{\partial x^2 \partial z}. \tag{15}$$

Equations (14) and (15) are usually called governing equations. Let us introduce the following dimensionless quantities:

$$X = \frac{x}{h}, Z = \frac{z}{h}, S_{13} = \frac{s_{13}}{s_{11}}, S_{33} = \frac{s_{33}}{s_{11}}, S_{44} = \frac{s_{44}}{s_{11}}, \overline{d}_{31} = \frac{d_{31}}{\sqrt{s_{11}\lambda_{11}}}, \overline{d}_{33} = \frac{d_{33}}{\sqrt{s_{11}\lambda_{11}}}, \overline{d}_{15} = \frac{d_{15}}{\sqrt{s_{11}\lambda_{11}}},$$

$$\Phi = \frac{\phi\sqrt{s_{11}\lambda_{11}}}{h}, \overline{\lambda}_{33} = \frac{\lambda_{33}}{\lambda_{11}}, \overline{P} = \frac{P}{h^2}s_{11}, \overline{b} = \frac{b}{h}, \overline{M} = \frac{M}{h^3}s_{11}, \overline{q} = qs_{11}, \overline{u} = \frac{u}{h}, \overline{w} = \frac{w}{h}, \overline{U} = \frac{Us_{11}}{h^2} \tag{16}$$

From Equation (16), Equations (14) and (15) can be transformed into

$$\overline{d}_{31}\frac{\partial^3 \overline{U}}{\partial Z^3} + \overline{d}_{33}\frac{\partial^3 \overline{U}}{\partial X^2 \partial Z} - \overline{d}_{15}\frac{\partial^3 \overline{U}}{\partial X^2 \partial Z} = \overline{\lambda}_{33}\frac{\partial^2 \Phi}{\partial Z^2} + \frac{\partial^2 \Phi}{\partial X^2} \tag{17}$$

and

$$\frac{\partial^4 \overline{U}}{\partial Z^4} + (2S_{13} + S_{44})\frac{\partial^4 \overline{U}}{\partial X^2 \partial Z^2} + S_{33}\frac{\partial^4 \overline{U}}{\partial X^4} = \overline{d}_{31}\frac{\partial^3 \Phi}{\partial Z^3} + \overline{d}_{33}\frac{\partial^3 \Phi}{\partial X^2 \partial Z} - \overline{d}_{15}\frac{\partial^3 \Phi}{\partial X^2 \partial Z}. \tag{18}$$

The boundary conditions can be transformed into

$$\int_{-1/2}^{1/2} \frac{\partial^2 \overline{U}}{\partial Z \partial X}dZ = -\frac{\overline{P}}{\overline{b}}, \int_{-1/2}^{1/2} \frac{\partial^2 \overline{U}}{\partial Z^2}dZ = 0, \int_{-1/2}^{1/2} Z\frac{\partial^2 \overline{U}}{\partial Z^2}dZ = \frac{\overline{M}}{\overline{b}}, \text{ at } X = 0, \tag{19}$$

$$\frac{\partial^2 \overline{U}}{\partial X^2} = -\frac{\partial^2 \overline{U}}{\partial X \partial Z} = 0, \text{ at } Z = 1/2, \tag{20}$$

$$\frac{\partial^2 \overline{U}}{\partial X^2} = -\overline{q}, -\frac{\partial^2 \overline{U}}{\partial X \partial Z} = 0, \text{ at } Z = -1/2, \tag{21}$$

$$\int_{-1/2}^{1/2} (-\overline{d}_{15}\frac{\partial^2 \overline{U}}{\partial Z \partial X} - \frac{\partial \Phi}{\partial X})dZ = 0, \text{ at } X = 0 \text{ and } X = l/h, \tag{22}$$

$$\frac{\partial \Phi}{\partial Z} = 0, \text{ at } Z = \pm 1/2, \tag{23}$$

and

$$\overline{u} = 0, \overline{w} = 0, \frac{\partial \overline{w}}{\partial X} = 0, \text{ at } Z = 0 \text{ and } X = l/h. \tag{24}$$

3. Multi-parameter Perturbation Solution

Equations (17) and (18) are two partial differential equations which are usually difficult to solve analytically. Here, we use the multi-parameter perturbation method to solve them. The piezoelectric coefficients are usually very small [33], thus, they can be selected as perturbation parameters to meet the requirement of convergence in perturbation expansions. From the point of view of the perturbation idea, if the cantilever beam without piezoelectric properties is regarded as an unperturbed system, the piezoelectric cantilever beam can be looked upon as a perturbed system. Selecting $\bar{d}_{31}$, $\bar{d}_{33}$ and $\bar{d}_{15}$ as the perturbation parameters, the Φ and $\overline{U}$ can be expanded as

$$
\begin{aligned}
\Phi = \Phi_0^0 + \Phi_1^I \bar{d}_{31} + \Phi_2^I \bar{d}_{33} + \Phi_3^I \bar{d}_{15} + \Phi_1^{II}(\bar{d}_{31})^2 + \Phi_2^{II}(\bar{d}_{33})^2 \\
+ \Phi_3^{II}(\bar{d}_{15})^2 + \Phi_4^{II}\bar{d}_{31}\bar{d}_{33} + \Phi_5^{II}\bar{d}_{31}\bar{d}_{15} + \Phi_6^{II}\bar{d}_{33}\bar{d}_{15}
\end{aligned}
\tag{25}
$$

and

$$
\begin{aligned}
\overline{U} = \overline{U}_0^0 + \overline{U}_1^I \bar{d}_{31} + \overline{U}_2^I \bar{d}_{33} + \overline{U}_3^I \bar{d}_{15} + \overline{U}_1^{II}(\bar{d}_{31})^2 + \overline{U}_2^{II}(\bar{d}_{33})^2 \\
+ \overline{U}_3^{II}(\bar{d}_{15})^2 + \overline{U}_4^{II}\bar{d}_{31}\bar{d}_{33} + \overline{U}_5^{II}\bar{d}_{31}\bar{d}_{15} + \overline{U}_6^{II}\bar{d}_{33}\bar{d}_{15}
\end{aligned}
\tag{26}
$$

where Φ_0^0 and $\overline{U}_0^0$, Φ_i^I and $\overline{U}_i^I$ ($i = 1, 2, 3$), and Φ_i^{II} and $\overline{U}_i^{II}$ ($i = 1, 2, \ldots, 5, 6$) are unknown functions of X and Z.

First, we solve the zero-order perturbation equations. Substituting Equations (25) and (26) into Equations (17) and (18) and comparing the coefficients of $(\bar{d}_{31})^0$, $(\bar{d}_{33})^0$ and $(\bar{d}_{15})^0$, we may obtain the zero-order perturbation equations

$$
\begin{cases}
\bar{\lambda}_{33}\dfrac{\partial^2 \Phi_0^0}{\partial Z^2} + \dfrac{\partial^2 \Phi_0^0}{\partial X^2} = 0 \\[2mm]
\dfrac{\partial^4 \overline{U}_0^0}{\partial Z^4} + (2S_{13} + S_{44})\dfrac{\partial^4 \overline{U}_0^0}{\partial X^2 \partial Z^2} + S_{33}\dfrac{\partial^4 \overline{U}_0^0}{\partial X^4} = 0
\end{cases}
\tag{27}
$$

The corresponding boundary conditions are

$$
\int_{-1/2}^{1/2}\left(-\frac{\partial^2 \overline{U}_0^0}{\partial Z \partial X}\right)dZ = -\frac{\overline{P}}{b}, \; \int_{-1/2}^{1/2}\frac{\partial^2 \overline{U}_0^0}{\partial Z^2}dZ = 0, \; \int_{-1/2}^{1/2} Z\frac{\partial^2 \overline{U}_0^0}{\partial Z^2}dZ = \frac{\overline{M}}{b}, \; \text{at } X = 0,
\tag{28}
$$

$$
\frac{\partial^2 \overline{U}_0^0}{\partial X^2} = -\frac{\partial^2 \overline{U}_0^0}{\partial X \partial Z} = 0, \text{ at } Z = 1/2,
\tag{29}
$$

$$
\frac{\partial^2 \overline{U}_0^0}{\partial X^2} = -\bar{q}, -\frac{\partial^2 \overline{U}_0^0}{\partial X \partial Z} = 0, \text{ at } Z = -1/2,
\tag{30}
$$

$$
\int_{-1/2}^{1/2}\left(-\frac{\partial \Phi_0^0}{\partial X}\right)dZ = 0, \text{ at } X = 0 \text{ and } X = l/h
\tag{31}
$$

and

$$
\frac{\partial \Phi_0^0}{\partial Z} = 0, \text{ at } Z = \pm 1/2.
\tag{32}
$$

Suppose,

$$
\begin{cases}
\Phi_0^0 = X^2 g_1^0(Z) + X g_2^0(Z) + g_3^0(Z) \\[2mm]
\overline{U}_0^0 = \frac{X^2}{2}f_1^0(Z) + X f_2^0(Z) + f_3^0(Z)
\end{cases}
\tag{33}
$$

where $g_i^0(Z)$ and $f_i^0(Z)$ ($i = 1, 2, 3$) are unknown functions of Z which can be determined by Equations (27) and (33), please see Appendix A.

Next, let us solve the first-order perturbation equations. Comparing the coefficients of $(\bar{d}_{31})^1$, $(\bar{d}_{33})^1$ and $(\bar{d}_{15})^1$, we may obtain the first-order perturbation equations as follows.

For term $(\bar{d}_{31})^1$:

$$
\begin{cases}
\dfrac{\partial^3 \overline{U}_0^0}{\partial Z^3} - \bar{\lambda}_{33}\dfrac{\partial^2 \Phi_1^I}{\partial Z^2} - \dfrac{\partial^2 \Phi_1^I}{\partial X^2} = 0 \\[2mm]
\dfrac{\partial^4 \overline{U}_1^I}{\partial Z^4} + (2S_{13} + S_{44})\dfrac{\partial^4 \overline{U}_1^I}{\partial X^2 \partial Z^2} + S_{33}\dfrac{\partial^4 \overline{U}_1^I}{\partial X^4} - \dfrac{\partial^3 \Phi_0^0}{\partial Z^3} = 0
\end{cases}
\tag{34}
$$

for term $(\bar{d}_{33})^1$:

$$\begin{cases} \dfrac{\partial^3 \overline{U}_0^0}{\partial X^2 \partial Z} = \overline{\lambda}_{33}\dfrac{\partial^2 \Phi_2^I}{\partial Z^2} + \dfrac{\partial^2 \Phi_2^I}{\partial X^2} \\ \dfrac{\partial^4 \overline{U}_2^I}{\partial Z^4} + (2S_{13} + S_{44})\dfrac{\partial^4 \overline{U}_2^I}{\partial X^2 \partial Z^2} + S_{33}\dfrac{\partial^4 \overline{U}_2^I}{\partial X^4} - \dfrac{\partial^3 \Phi_0^I}{\partial X^2 \partial Z} = 0 \end{cases} \tag{35}$$

and for term $(\bar{d}_{15})^1$:

$$\begin{cases} -\dfrac{\partial^3 \overline{U}_0^0}{\partial X^2 \partial Z} = \overline{\lambda}_{33}\dfrac{\partial^2 \Phi_3^I}{\partial Z^2} + \dfrac{\partial^2 \Phi_3^I}{\partial X^2} \\ \dfrac{\partial^4 \overline{U}_3^I}{\partial Z^4} + (2S_{13} + S_{44})\dfrac{\partial^4 \overline{U}_3^I}{\partial X^2 \partial Z^2} + S_{33}\dfrac{\partial^4 \overline{U}_3^I}{\partial X^4} + \dfrac{\partial^3 \Phi_0^I}{\partial X^2 \partial Z} = 0 \end{cases} \tag{36}$$

The corresponding boundary conditions are

$$\begin{cases} \int_{-1/2}^{1/2} -\dfrac{\partial^2 \overline{U}_1^I}{\partial Z \partial X} dZ = 0, \\ \int_{-1/2}^{1/2} \dfrac{\partial^2 \overline{U}_2^I}{\partial Z^2} dZ = 0, \quad , \text{ at } X = 0, \\ \int_{-1/2}^{1/2} Z\dfrac{\partial^2 \overline{U}_3^I}{\partial Z^2} dZ = 0 \end{cases} \tag{37}$$

$$\dfrac{\partial^2 \overline{U}_i^I}{\partial X^2} = -\dfrac{\partial^2 \overline{U}_i^I}{\partial X \partial Z} = 0 (i = 1,2,3), \text{ at } Z = 1/2, \tag{38}$$

$$\dfrac{\partial^2 \overline{U}_i^I}{\partial X^2} = 0, -\dfrac{\partial^2 \overline{U}_i^I}{\partial X \partial Z} = 0 (i = 1,2,3), \text{ at } Z = -1/2, \tag{39}$$

$$\begin{cases} \int_{-1/2}^{1/2} \left(-\dfrac{\partial \Phi_1^I}{\partial X}\right) dZ = 0 \\ \int_{-1/2}^{1/2} \left(-\dfrac{\partial \Phi_2^I}{\partial X}\right) dZ = 0 \quad , \text{ at } X = 0 \text{ and } X = l/h, \\ \int_{-1/2}^{1/2} \left(-\dfrac{\partial^2 \overline{U}_0^0}{\partial Z \partial X} - \dfrac{\partial \Phi_3^I}{\partial X}\right) dZ = 0 \end{cases} \tag{40}$$

and

$$\dfrac{\partial \Phi_i^I}{\partial Z} = 0 (i = 1,2,3), \text{ at } Z = \pm 1/2. \tag{41}$$

Similarly, suppose

$$\begin{cases} \Phi_i^I = X^2 g_{3i-2}^I(Z) + X g_{3i-1}^I(Z) + g_{3i}^I(Z) \\ \overline{U}_i^I = \dfrac{X^2}{2} f_{3i-2}^I(Z) + X f_{3i-1}^I(Z) + f_{3i}^I(Z) \end{cases} (i = 1,2,3), \tag{42}$$

where $g_i^I(Z)$ and $f_i^I(Z)$ $(i = 1,2,3,\ldots,9)$ are unknown functions of Z which can be determined by Equations (34)–(36) and (42), please see Appendix A.

Then, we solve the second-order perturbation equations. Comparing the coefficients of $(d_{31})^2$, $(d_{33})^2$, $(d_{15})^2$, $d_{31}d_{33}$, $d_{31}d_{15}$ and $d_{33}d_{15}$, we may obtain the two-order perturbation equations as follows.

For term $(d_{31})^2$:

$$\begin{cases} \dfrac{\partial^3 \overline{U}_1^I}{\partial Z^3} = \overline{\lambda}_{33}\dfrac{\partial^2 \Phi_1^{II}}{\partial Z^2} + \dfrac{\partial^2 \Phi_1^{II}}{\partial X^2} \\ \dfrac{\partial^4 \overline{U}_1^{II}}{\partial Z^4} + (2S_{13} + S_{44})\dfrac{\partial^4 \overline{U}_1^{II}}{\partial X^2 \partial Z^2} + S_{33}\dfrac{\partial^4 \overline{U}_1^{II}}{\partial X^4} = \dfrac{\partial^3 \Phi_1^I}{\partial Z^3} \end{cases} \tag{43}$$

for term $(d_{33})^2$:

$$\begin{cases} \dfrac{\partial^3 \overline{U}_2^I}{\partial X^2 \partial Z} = \overline{\lambda}_{33}\dfrac{\partial^2 \Phi_2^{II}}{\partial Z^2} + \dfrac{\partial^2 \Phi_2^{II}}{\partial X^2} \\ \dfrac{\partial^4 \overline{U}_2^{II}}{\partial Z^4} + (2S_{13} + S_{44})\dfrac{\partial^4 \overline{U}_2^{II}}{\partial X^2 \partial Z^2} + S_{33}\dfrac{\partial^4 \overline{U}_2^{II}}{\partial X^4} = \dfrac{\partial^3 \Phi_2^I}{\partial X^2 \partial Z} \end{cases} \tag{44}$$

for term $(d_{15})^2$:

$$\begin{cases} -\dfrac{\partial^3 \overline{U}_3^I}{\partial X^2 \partial Z} = \overline{\lambda}_{33}\dfrac{\partial^2 \Phi_3^{II}}{\partial Z^2} + \dfrac{\partial^2 \Phi_3^{II}}{\partial X^2} \\[2mm] \dfrac{\partial^4 \overline{U}_3^{II}}{\partial Z^4} + (2S_{13} + S_{44})\dfrac{\partial^4 \overline{U}_3^{II}}{\partial X^2 \partial Z^2} + S_{33}\dfrac{\partial^4 \overline{U}_3^{II}}{\partial X^4} = -\dfrac{\partial^3 \Phi_3^I}{\partial X^2 \partial Z} \end{cases} \tag{45}$$

for term $d_{31}d_{33}$:

$$\begin{cases} \dfrac{\partial^3 \overline{U}_2^I}{\partial Z^3} + \dfrac{\partial^3 \overline{U}_1^I}{\partial X^2 \partial Z} = \overline{\lambda}_{33}\dfrac{\partial^2 \Phi_4^{II}}{\partial Z^2} + \dfrac{\partial^2 \Phi_4^{II}}{\partial X^2} \\[2mm] \dfrac{\partial^4 \overline{U}_4^{II}}{\partial Z^4} + (2S_{13} + S_{44})\dfrac{\partial^4 \overline{U}_4^{II}}{\partial X^2 \partial Z^2} + S_{33}\dfrac{\partial^4 \overline{U}_4^{II}}{\partial X^4} = \dfrac{\partial^3 \Phi_2^I}{\partial Z^3} + \dfrac{\partial^3 \Phi_1^I}{\partial X^2 \partial Z} \end{cases} \tag{46}$$

for term $d_{31}d_{15}$:

$$\begin{cases} \dfrac{\partial^3 \overline{U}_3^I}{\partial Z^3} - \dfrac{\partial^3 \overline{U}_1^I}{\partial X^2 \partial Z} = \overline{\lambda}_{33}\dfrac{\partial^2 \Phi_5^{II}}{\partial Z^2} + \dfrac{\partial^2 \Phi_5^{II}}{\partial X^2} \\[2mm] \dfrac{\partial^4 \overline{U}_5^{II}}{\partial Z^4} + (2S_{13} + S_{44})\dfrac{\partial^4 \overline{U}_5^{II}}{\partial X^2 \partial Z^2} + S_{33}\dfrac{\partial^4 \overline{U}_5^{II}}{\partial X^4} = \dfrac{\partial^3 \Phi_3^I}{\partial Z^3} - \dfrac{\partial^3 \Phi_1^I}{\partial X^2 \partial Z} \end{cases} \tag{47}$$

and for term $d_{33}d_{15}$:

$$\begin{cases} \dfrac{\partial^3 \overline{U}_3^I}{\partial X^2 \partial Z} - \dfrac{\partial^3 \overline{U}_2^I}{\partial X^2 \partial Z} = \overline{\lambda}_{33}\dfrac{\partial^2 \Phi_6^{II}}{\partial Z^2} + \dfrac{\partial^2 \Phi_6^{II}}{\partial X^2} \\[2mm] \dfrac{\partial^4 \overline{U}_6^{II}}{\partial Z^4} + (2S_{13} + S_{44})\dfrac{\partial^4 \overline{U}_6^{II}}{\partial X^2 \partial Z^2} + S_{33}\dfrac{\partial^4 \overline{U}_6^{II}}{\partial X^4} = \dfrac{\partial^3 \Phi_3^I}{\partial X^2 \partial Z} - \dfrac{\partial^3 \Phi_2^I}{\partial X^2 \partial Z} \end{cases} \tag{48}$$

The corresponding boundary conditions are

$$\begin{cases} \displaystyle\int_{-1/2}^{1/2} \dfrac{\partial^2 \overline{U}_1^{II}}{\partial Z \partial X}dZ = 0, \\[3mm] \displaystyle\int_{-1/2}^{1/2} \dfrac{\partial^2 \overline{U}_2^{II}}{\partial Z^2}dZ = 0, \quad , \text{ at } X = 0, \\[3mm] \displaystyle\int_{-1/2}^{1/2} Z\dfrac{\partial^2 \overline{U}_3^{II}}{\partial Z^2}dZ = 0 \end{cases} \tag{49}$$

$$\dfrac{\partial^2 \overline{U}_i^{II}}{\partial X^2} = -\dfrac{\partial^2 \overline{U}_i^{II}}{\partial X \partial Z} = 0 (i = 1, 2, 3, 4, 5, 6), \text{ at } Z = 1/2, \tag{50}$$

$$\dfrac{\partial^2 \overline{U}_i^{II}}{\partial X^2} = 0, -\dfrac{\partial^2 \overline{U}_i^{II}}{\partial X \partial Z} = 0 (i = 1, 2, 3, 4, 5, 6), \text{ at } Z = -1/2, \tag{51}$$

$$\begin{cases} \displaystyle\int_{-1/2}^{1/2} \dfrac{\partial \Phi_1^{II}}{\partial X}dZ = 0, \displaystyle\int_{-1/2}^{1/2} \dfrac{\partial \Phi_2^{II}}{\partial X}dZ = 0 \\[3mm] \displaystyle\int_{-1/2}^{1/2} \left(\dfrac{\partial^2 \overline{U}_3^I}{\partial X \partial Z} + \dfrac{\partial \Phi_3^{II}}{\partial X}\right)dZ = 0, \displaystyle\int_{-1/2}^{1/2} \dfrac{\partial \Phi_4^{II}}{\partial X}dZ = 0 \quad , \text{ at } X = 0 \text{ and } X = l/h, \\[3mm] \displaystyle\int_{-1/2}^{1/2} \left(\dfrac{\partial^2 \overline{U}_1^I}{\partial X \partial Z} + \dfrac{\partial \Phi_5^{II}}{\partial X}\right)dZ = 0, \displaystyle\int_{-1/2}^{1/2} \left(\dfrac{\partial^2 \overline{U}_2^I}{\partial X \partial Z} + \dfrac{\partial \Phi_6^{II}}{\partial X}\right)dZ = 0 \end{cases} \tag{52}$$

and

$$\dfrac{\partial \Phi_i^{II}}{\partial Z} = 0, \text{ at } Z = \pm 1/2. \tag{53}$$

Suppose,

$$\begin{cases} \Phi_i^{II} = X^2 g_{3i-2}^{II}(Z) + X g_{3i-1}^{II}(Z) + g_{3i}^{II}(Z) \\[2mm] \overline{U}_i^{II} = \dfrac{X^2}{2}f_{3i-2}^{II}(Z) + X f_{3i-1}^{II}(Z) + f_{3i}^{II}(Z) \end{cases} (i = 1, 2, 3, 4, 5, 6), \tag{54}$$

where $g_i^{II}(Z)$ and $f_i^{II}(Z)$ $(i = 1, 2, 3, \ldots, 18)$ are unknown functions of Z which can be determined by Equations (43)–(48) and (54), please see Appendix A.

Thus, we can obtain

$$\Phi = B_6^0 + \bar{d}_{31}\left[X^2\left(-\frac{3}{\bar{\lambda}_{33}}\bar{q}Z^2 + \frac{1}{4\bar{\lambda}_{33}}\bar{q}\right) + X\left(-\frac{6}{b\bar{\lambda}_{33}}\bar{P}Z^2 + \frac{1}{2b\bar{\lambda}_{33}}\bar{P}\right) + \frac{1}{2(\bar{\lambda}_{33})^2}\bar{q}Z^4 - \frac{1}{4(\bar{\lambda}_{33})^2}\bar{q}Z^2 \right.$$
$$+ \frac{(2S_{13}+S_{44})}{2\bar{\lambda}_{33}}\bar{q}Z^4 + \frac{6}{b\bar{\lambda}_{33}}\bar{M}Z^2 - \frac{3(2S_{13}+S_{44})}{20\bar{\lambda}_{33}}\bar{q}Z^2 + B_6^I\right] + \bar{d}_{33}\left[-\frac{1}{2\bar{\lambda}_{33}}\bar{q}Z^4 + \frac{3}{4\bar{\lambda}_{33}}\bar{q}Z^2 \right.$$
$$\left. - \frac{1}{2\bar{\lambda}_{33}}\bar{q}Z + B_{12}^I\right] + \bar{d}_{15}\left[-\frac{1}{2}\bar{q}X^2 - \frac{\bar{P}}{b}X + \frac{1}{2\bar{\lambda}_{33}}\bar{q}Z^2 + \frac{1}{2\bar{\lambda}_{33}}\bar{q}Z^4 - \frac{3}{4\bar{\lambda}_{33}}\bar{q}Z^2 + B_{18}^I\right]$$
$$+ (\bar{d}_{31})^2 B_6^{II} + (\bar{d}_{33})^2 B_{12}^{II} + (\bar{d}_{15})^2 B_{18}^{II} + \bar{d}_{31}\bar{d}_{33}B_{24}^{II} + \bar{d}_{31}\bar{d}_{15}B_{30}^{II} + \bar{d}_{33}\bar{d}_{15}B_{36}^{II} \tag{55}$$

and

$$\bar{U} = \frac{X^2}{2}\left(-2\bar{q}Z^3 + \frac{3}{2}\bar{q}Z - \frac{\bar{q}}{2}\right) + X\left(-\frac{2}{b}\bar{P}Z^3 + \frac{3}{2b}\bar{P}Z + C_8^0\right) + \frac{(2S_{13}+S_{44})}{10}\bar{q}Z^5 + \frac{2}{b}\bar{M}Z^3$$
$$- \frac{(2S_{13}+S_{44})}{20}\bar{q}Z^3 + C_{11}^0 Z + C_{12}^0 + \bar{d}_{31}\left(XC_8^I + C_{11}^I Z + C_{12}^I\right) + \bar{d}_{33}\left(XC_{20}^I + C_{23}^I Z + C_{24}^I\right)$$
$$+ \bar{d}_{15}\left(XC_{32}^I + C_{35}^I Z + C_{36}^I\right) + (\bar{d}_{31})^2\left[XC_8^{II} + \frac{1}{10(\bar{\lambda}_{33})^2}\bar{q}Z^5 + \frac{(2S_{13}+S_{44})}{10\bar{\lambda}_{33}}\bar{q}Z^5 - \frac{1}{20(\bar{\lambda}_{33})^2}\bar{q}Z^3\right.$$
$$\left. - \frac{(2S_{13}+S_{44})}{20\bar{\lambda}_{33}}\bar{q}Z^3 + C_{11}^{II} Z + C_{12}^{II}\right] + (\bar{d}_{33})^2\left(XC_{20}^{II} + C_{23}^{II} Z + C_{24}^{II}\right) + (\bar{d}_{15})^2\left(XC_{32}^{II} + C_{35}^{II} Z + C_{36}^{II}\right)$$
$$+ \bar{d}_{31}\bar{d}_{33}\left(XC_{44}^{II} - \frac{1}{5\bar{\lambda}_{33}}\bar{q}Z^5 + \frac{1}{10\bar{\lambda}_{33}}\bar{q}Z^3 + C_{47}^{II} Z + C_{48}^{II}\right) + \bar{d}_{31}\bar{d}_{15}\left(XC_{56}^{II} + \frac{1}{5\bar{\lambda}_{33}}\bar{q}Z^5 - \frac{1}{10\bar{\lambda}_{33}}\bar{q}Z^3\right.$$
$$\left. + C_{59}^{II} Z + C_{60}^{II}\right) + \bar{d}_{33}\bar{d}_{15}\left(XC_{68}^{II} + C_{71}^{II} Z + C_{72}^{II}\right) \tag{56}$$

Finally, from Equations (55) and (56), we can obtain the expression of displacement components, stress components, and electric displacement components. The detailed derivation is shown in Appendix B. Thus, the bending problem of a piezoelectric cantilever beam under combined loads is solved. It can be seen from the derivation above that the piezoelectric effect is not shown in the zero-order perturbation solution, that is, the zero-order perturbation solution is the solution of the cantilever beam without piezoelectric properties which is regarded as the unperturbed system. The piezoelectric properties are only shown in the first-order and second-order perturbation solutions. In other words, the mechanical meaning of the first-order and second-order solutions is the influence of piezoelectric properties on the deformation of piezoelectric cantilever beams. This phenomenon is consistent with the basic idea of perturbation method.

4. Comparison of the Solution Presented Here and the Existing Solution

The theoretical solution for a piezoelectric cantilever beam under combined loads is given in this paper by a new method which is usually called the multi-parameter perturbation method. The validity of the theoretical solution should further be verified. For this purpose, we compare the solution presented here with the solution given in reference [17].

Before the comparison, we need to make a degradation of the solution presented here. In reference [17], only the concentrated force is considered. In this paper, however, the concentrated force, bending moment, and uniformly distributed load are all considered. Thus, for the convenience of comparison, we let the bending moment and uniformly distributed load equal to zero, that is, let

$$q = 0, M = 0. \tag{57}$$

Substituting Equation (57) into Equations (A44) and (A45), the displacement components can be transformed into

$$w = \left(\frac{d_{33}d_{31}}{\lambda_{33}} - s_{13}\right)\frac{6P}{bh^3}xz^2 - \left(\frac{d_{31}d_{31}}{\lambda_{33}} - s_{11}\right)\frac{2P}{bh^3}x^3$$
$$+ \left(\frac{d_{31}d_{31}}{\lambda_{33}} - s_{11}\right)\frac{6P}{bh^3}l^2 x - \left(\frac{d_{31}d_{31}}{\lambda_{33}} - s_{11}\right)\frac{4P}{bh^3}l^3 \tag{58}$$

and

$$u = \left(\frac{d_{31}d_{31}}{\lambda_{33}} - s_{11}\right)\frac{6P}{bh^3}x^2z - \left(\frac{d_{33}d_{31}}{\lambda_{33}} - s_{13} - s_{44} - \frac{d_{15}d_{31}}{\lambda_{33}}\right)\frac{2P}{bh^3}z^3$$
$$-\left(\frac{d_{31}d_{31}}{\lambda_{33}} - s_{11}\right)\frac{6P}{bh^3}l^2z - \left(\frac{d_{15}d_{31}}{\lambda_{33}} + s_{44}\right)\frac{3P}{2bh}z \tag{59}$$

Similarly, substituting Equation (57) into Equations (A46), (A47), and (A48), the stress components can be written as

$$\sigma_x = -\frac{12P}{bh^3}xz, \tag{60}$$

$$\sigma_z = 0, \tag{61}$$

and

$$\tau_{zx} = \frac{6P}{bh^3}z^2 - \frac{3P}{2bh}. \tag{62}$$

Substituting Equation (57) into Equations (A49) and (A50), the expressions of electric displacement components are

$$D_x = (d_{15} + \frac{\lambda_{11}d_{31}}{\lambda_{33}})(\frac{6P}{bh^3}z^2 - \frac{3P}{2bh}) \tag{63}$$

and

$$D_z = 0. \tag{64}$$

By comparing Equations (58)–(64) with the expressions of displacement components, stress components, and electric displacement components in reference [17], it can be found that they are exactly the same, which indicates that the solution obtained here is correct. It should be mentioned that the structures studied in this paper and in reference [17] are both piezoelectric cantilever beams, but the structure in this paper is subjected to combined loads and the structure in reference [17] is subjected only to a concentrated force. In addition, non-dimensionalization is considered, the relevant experiments are carried out, and a new method called the multi-parameter perturbation method is given in this paper. These differences mentioned above constitute the advancements of this paper, compared with reference [17].

5. Experimental Verification

To further verify the validity of the theoretical solution presented here, we carry out the relevant experiments of piezoelectric cantilever beams. The mechanical model of the theoretical part is shown in Figure 1, it can be seen that it is a piezoelectric cantilever beam subjected to three kinds of loads. In the experiment, it is very difficult to apply these three kinds of loads at the same time. Therefore, we apply only the concentrated force at the cantilever end to carry out the experiments, that is, this experiment corresponds only to the case where the bending moment and the uniformly distributed load in the theoretical solution are zero. The details of the experiments are as follows. The main experimental equipments include a non-contact laser displacement sensor (ZSY Group Ltd, London, UK), a bench clamp (a cantilever beam clamping device), weights, and the ZLDS10X measuring software (ZSY Group Ltd, London, UK). The measuring range of the non-contact laser displacement sensor is 1 m, the accuracy is 0.01%, and the sampling frequency is 2 kHz. The experimental specimens consist of two groups of PbZrTiO$_3$-5 (Generally abbreviated as PZT-5) piezoelectric ceramic sheets in which one group has piezoelectric properties and the other group has no piezoelectric properties. The size of the experimental specimens is 60 mm × 10 mm × 1 mm. The experimental specimen and non-contact laser displacement sensor are shown in Figure 2, the experimental device is shown in Figure 3, and the material constants are shown in Table 1.

The clamping length of the experimental specimens is 10 mm, therefore, the length of the piezoelectric cantilever beam is 50 mm. The deformations of the free end of piezoelectric cantilever beam are measured at the applied load 0.49 N, 0.98 N, and 1.96 N. The measured experimental data and theoretical calculation results are shown in Tables 2 and 3, respectively. It should be noted that the self-weight of the piezoelectric cantilever beam is 0.0367 N, and the ratio of the self-weight to the minimum applied load is 0.075, which indicates that the self-weight of the piezoelectric cantilever beam is very small and thus may be ignored.

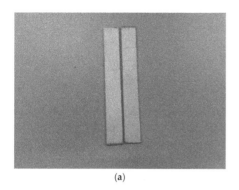

(a)

(b)

Figure 2. Scheme of experimental specimens and measuring instruments: (**a**) PbZrTiO₃-5 (Generally abbreviated as PZT-5) piezoelectric ceramic specimens. (**b**) The non-contact laser displacement sensor.

(a)

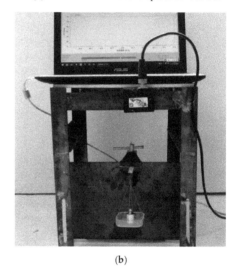

(b)

Figure 3. Scheme of experimental device: (**a**) The cantilever beam device. (**b**) The integral measuring device.

Table 1. Physical properties of PZT-5 materials [33].

Elastic Constant (10^{-12} m²·N⁻¹)					Piezoelectric Constant (10^{-12} C·N⁻¹)			Dielectric Constant (10^{-8} F·m⁻¹)	
s^0_{11}	s^0_{12}	s^0_{13}	s^0_{33}	s^0_{44}	d^0_{31}	d^0_{33}	d^0_{15}	λ^0_{11}	λ^0_{33}
16.4	−5.74	−7.22	18.8	47.5	−172	374	584	1.505	1.531

Table 2. Comparison of experimental data and theoretical calculation results.

Loads(N)	The Deformation of the Cantilever End		
	Experimental Data (mm)	**Theoretical Results (mm)**	**Relative Errors (%)**
0.49	0.4069	0.3545	12.87
0.98	0.7527	0.7089	5.82
1.96	1.6072	1.4178	11.79

Polymers **2019**, *11*, 1934

Table 3. Comparison of deformation test results between piezoelectric cantilever beam and cantilever beam without piezoelectric properties.

Loads(N)	The Deformation of the Cantilever End		
	Piezoelectric Cantilever Beam (mm)	Cantilever Beam without Piezoelectric Properties (mm)	Difference (mm)
0.49	0.4069	0.5351	0.1282
0.98	0.7527	0.8463	0.0936
1.96	1.6072	1.9796	0.3724

From Table 2, it can be seen that the theoretical results are in good agreement with the experimental results, and the relative errors under every level load are less than 15% allowed in engineering. This indicates that the analytical solution presented in this paper is reliable.

Table 3 shows that the deformation of the piezoelectric cantilever beam is smaller than the cantilever beam without piezoelectric properties. This means that the piezoelectric properties have a certain effect on the deformation of the piezoelectric cantilever beam, and its effect is, to a certain extent, hindering the deformation of the cantilever beam. This phenomenon can be explained by energy conservation. For piezoelectric cantilever beams, part of the work done by external forces is transformed into the elastic strain energy of piezoelectric cantilever beams, while the other part is transformed into the electric energy due to the existence of piezoelectric properties. For cantilever beams without piezoelectric properties, the work done by external forces is basically transformed into the elastic strain energy of cantilever beams. Therefore, the deformation of cantilever beams without piezoelectric properties is larger than that of cantilever beams with piezoelectric properties. The phenomenon mentioned above is commonly known as the piezoelectric stiffening effect peculiar to piezoelectric materials and structures.

6. Conclusions

In this study, we used a multi-parameter perturbation method to solve the bending deformation problem of piezoelectric cantilever beams under combined loads. And we compared the solution presented here with the existing solution from Yang and Liu [17] to validate the rationality of the presented solution. In addition, we carried out the related experiments of the piezoelectric cantilever beam, and compared the experimental results with the theoretical solution presented here, and also investigated the influence of the piezoelectric properties on the deformation of piezoelectric cantilever beams. The following main conclusions can be drawn.

(i) The theoretical results are in good agreement with the experimental results, which means that the analytical solution given in this paper is correct and the multi-parameter perturbation method is effective.

(ii) From the perturbation expansion, it is easy to find that the zero-order perturbation solution is a pure mechanical solution, in which the piezoelectric effect has not been incorporated. From the first-order, second-order, and higher order perturbation solutions, the piezoelectric effect is gradually reflected. This structural form of the multi-parameter perturbation solution presented here is beneficial to the analysis and understanding of the solved problem.

(iii) The deformation magnitude of a piezoelectric cantilever beam is smaller than that of a cantilever beam without piezoelectricity, due to the well-known piezoelectric stiffening effect.

Unfortunately, the numerical simulation for the physical system studied here has not been carried out in this study. In our previous study [34], we used ABAQUS software to simulate the problem of functionally graded piezoelectric cantilever beams with different properties in tension and compression. Similarly, the problem studied here may also be simulated by ABAQUS, which is our follow-up research. In summary, the multi-parameter perturbation method presented in this paper provides a new way to solve complex non-linear structural problems. The analytical solution of the bending problems of piezoelectric cantilever beams under combined loads can provide a theoretical basis and reference for the analysis and design of sensors or actuators made of piezoelectric polymers.

Author Contributions: Conceptualization, X.-T.H. and J.-Y.S.; funding acquisition, X.-T.H. and J.-Y.S.; methodology, X.-T.H. and Z.-X.Y.; data curation, Z.-X.Y. and H.-X.J.; writing—original draft preparation, X.-T.H. and Z.-X.Y.; writing—review and editing, H.-X.J. and J.-Y.S.

Funding: This project is supported by National Natural Science Foundation of China (Grant No. 11572061 and 11772072).

Conflicts of Interest: The authors declare no conflicts of interest.

Appendix A

(1) The unknown functions $g_i^0(Z)$ and $f_i^0(Z)$ ($i = 1, 2, 3$):

From Equations (27) and (33), we can obtain the unknown functions $g_i^0(Z)$ and $f_i^0(Z)$ ($i = 1, 2, 3$),

$$
\begin{cases}
g_1^0(Z) = B_1^0 Z + B_2^0 \\
g_2^0(Z) = B_3^0 Z + B_4^0 \\
g_3^0(Z) = -\frac{1}{3\bar{\lambda}_{33}} B_1^0 Z^3 - \frac{1}{\bar{\lambda}_{33}} B_2^0 Z^2 + B_5^0 Z + B_6^0
\end{cases}
\tag{A1}
$$

and

$$
\begin{cases}
f_1^0(Z) = \frac{1}{6} C_1^0 Z^3 + \frac{1}{2} C_2^0 Z^2 + C_3^0 Z + C_4^0 \\
f_2^0(Z) = \frac{1}{6} C_5^0 Z^3 + \frac{1}{2} C_6^0 Z^2 + C_7^0 Z + C_8^0 \\
f_3^0(Z) = -\frac{1}{120}(2S_{13} + S_{44}) C_1^0 Z^5 - \frac{1}{24}(2S_{13} + S_{44}) C_2^0 Z^4 \\
\quad + \frac{1}{6} C_9^0 Z^3 + \frac{1}{2} C_{10}^0 Z^2 + C_{11}^0 Z + C_{12}^0
\end{cases}
\tag{A2}
$$

where B_i^0 ($i = 1, 2, 3, \ldots, 6$) and C_i^0 ($i = 1, 2, 3, \ldots, 12$) are undetermined constants which can be determined by Equations (28)–(32),

$$
\begin{aligned}
&C_1^0 = -12\bar{q},\, C_2^0 = 0,\, C_3^0 = \frac{3}{2}\bar{q},\, C_4^0 = -\frac{\bar{q}}{2},\, C_5^0 = -\frac{12\bar{P}}{\bar{b}},\, C_6^0 = 0, \\
&C_7^0 = \frac{3}{2}\frac{\bar{P}}{\bar{b}},\, C_9^0 = \frac{12\bar{M}}{\bar{b}} - \frac{3}{10}(2S_{13} + S_{44})\bar{q},\, C_{10}^0 = 0
\end{aligned}
\tag{A3}
$$

$$
B_1^0 = 0,\, B_2^0 = 0,\, B_3^0 = 0,\, B_5^0 = 0.
\tag{A4}
$$

(2) The unknown functions $g_i^I(Z)$ and $f_i^I(Z)$ ($i = 1, 2, 3, \ldots, 9$):

From Equations (34)–(36) and (42), we can obtain the unknown functions $g_i^I(Z)$ and $f_i^I(Z)$ ($i = 1, 2, 3, \ldots, 9$),

$$
\begin{cases}
g_1^I(Z) = \frac{1}{4\bar{\lambda}_{33}} C_1^0 Z^2 + B_1^I Z + B_2^I \\
g_2^I(Z) = \frac{1}{2\bar{\lambda}_{33}} C_5^0 Z^2 + B_3^I Z + B_4^I \\
g_3^I(Z) = -\frac{1}{24(\bar{\lambda}_{33})^2} C_1^0 Z^4 - \frac{1}{3\bar{\lambda}_{33}} B_1^I Z^3 - \frac{1}{\bar{\lambda}_{33}} B_2^I Z^2 - \frac{(2S_{13}+S_{44})}{24\bar{\lambda}_{33}} C_1^0 Z^4 \\
\quad - \frac{(2S_{13}+S_{44})}{6\bar{\lambda}_{33}} C_2^0 Z^3 + \frac{1}{2\bar{\lambda}_{33}} C_9^0 Z^2 + B_5^I Z + B_6^I \\
g_6^I(Z) = -\frac{1}{3\bar{\lambda}_{33}} B_7^I Z^3 - \frac{1}{\bar{\lambda}_{33}} B_8^I Z^2 + \frac{C_1^0}{24\bar{\lambda}_{33}} Z^4 \\
\quad + \frac{C_2^0}{6\bar{\lambda}_{33}} Z^3 + \frac{C_3^0}{2\bar{\lambda}_{33}} Z^2 + B_{11}^I Z + B_{12}^I \\
g_9^I(Z) = -\frac{1}{3\bar{\lambda}_{33}} B_{13}^I Z^3 - \frac{1}{\bar{\lambda}_{33}} B_{14}^I Z^2 - \frac{C_1^0}{24\bar{\lambda}_{33}} Z^4 - \frac{C_2^0}{6\bar{\lambda}_{33}} Z^3 - \frac{C_3^0}{2\bar{\lambda}_{33}} Z^2 + B_{17}^I Z + B_{18}^I
\end{cases}
\tag{A5}
$$

$$
g_i^I(Z) = B_{2i-1}^I Z + B_{2i}^I\ (i = 4, 5, 7, 8),
\tag{A6}
$$

$$
f_i^I(Z) = \frac{1}{6} C_{4i-3}^I Z^3 + \frac{1}{2} C_{4i-2}^I Z^2 + C_{4i-1}^I Z + C_{4i}^I\ (i = 1, 2, 4, 5, 7, 8),
\tag{A7}
$$

and

$$
\begin{cases}
f_3^{\mathrm{I}}(Z) = -\frac{(2S_{13}+S_{44})}{120}C_1^{\mathrm{I}}Z^5 - \frac{(2S_{13}+S_{44})}{24}C_2^{\mathrm{I}}Z^4 - \frac{1}{12\bar{\lambda}_{33}}B_1^0 Z^4 \\
\quad + \frac{1}{6}C_9^{\mathrm{I}}Z^3 + \frac{1}{2}C_{10}^{\mathrm{I}}Z^2 + C_{11}^{\mathrm{I}}Z + C_{12}^{\mathrm{I}} \\
f_6^{\mathrm{I}}(Z) = -\frac{(2S_{13}+S_{44})}{120}C_{13}^{\mathrm{I}}Z^5 - \frac{(2S_{13}+S_{44})}{24}C_{14}^{\mathrm{I}}Z^4 + \frac{1}{12}B_1^0 Z^4 \\
\quad + \frac{1}{6}C_{21}^{\mathrm{I}}Z^3 + \frac{1}{2}C_{22}^{\mathrm{I}}Z^2 + C_{23}^{\mathrm{I}}Z + C_{24}^{\mathrm{I}} \\
f_9^{\mathrm{I}}(Z) = -\frac{(2S_{13}+S_{44})}{120}C_{25}^{\mathrm{I}}Z^5 - \frac{(2S_{13}+S_{44})}{24}C_{26}^{\mathrm{I}}Z^4 - \frac{B_1^0}{12}Z^4 \\
\quad + \frac{1}{6}C_{33}^{\mathrm{I}}Z^3 + \frac{1}{2}C_{34}^{\mathrm{I}}Z^2 + C_{35}^{\mathrm{I}}Z + C_{36}^{\mathrm{I}}
\end{cases} \tag{A8}
$$

where B_i^{I} $(i = 1,2,3,\ldots,18)$ and C_i^{I} $(i = 1,2,3,\ldots,36)$ are undetermined constants which can be determined by Equations (37)–(41),

$$
C_1^{\mathrm{I}} = 0, C_2^{\mathrm{I}} = 0, C_3^{\mathrm{I}} = 0, C_4^{\mathrm{I}} = 0, C_5^{\mathrm{I}} = 0, C_6^{\mathrm{I}} = 0, C_7^{\mathrm{I}} = 0, C_9^{\mathrm{I}} = 0, C_{10}^{\mathrm{I}} = \frac{1}{12\bar{\lambda}_{33}}B_1^0, \tag{A9}
$$

$$
B_1^{\mathrm{I}} = \frac{C_2^0}{2\bar{\lambda}_{33}}, B_2^{\mathrm{I}} = -\frac{C_1^0}{48\bar{\lambda}_{33}}, B_3^{\mathrm{I}} = \frac{C_6^0}{\bar{\lambda}_{33}}, B_4^{\mathrm{I}} = -\frac{C_5^0}{24\bar{\lambda}_{33}}, B_5^{\mathrm{I}} = \frac{C_{10}^0}{\bar{\lambda}_{33}} + \frac{C_2^0}{8(\bar{\lambda}_{33})^2}, \tag{A10}
$$

$$
C_{13}^{\mathrm{I}} = 0, C_{14}^{\mathrm{I}} = 0, C_{15}^{\mathrm{I}} = 0, C_{16}^{\mathrm{I}} = 0, C_{17}^{\mathrm{I}} = 0, C_{18}^{\mathrm{I}} = 0, C_{19}^{\mathrm{I}} = 0, C_{21}^{\mathrm{I}} = 0, C_{22}^{\mathrm{I}} = -\frac{1}{12}B_1^0, \tag{A11}
$$

$$
B_7^{\mathrm{I}} = 0, B_8^{\mathrm{I}} = 0, B_9^{\mathrm{I}} = 0, B_{10}^{\mathrm{I}} = 0, B_{11}^{\mathrm{I}} = \frac{C_4^0}{\bar{\lambda}_{33}}, \tag{A12}
$$

$$
C_{25}^{\mathrm{I}} = 0, C_{26}^{\mathrm{I}} = 0, C_{27}^{\mathrm{I}} = 0, C_{28}^{\mathrm{I}} = 0, C_{29}^{\mathrm{I}} = 0, C_{30}^{\mathrm{I}} = 0, C_{31}^{\mathrm{I}} = 0, C_{33}^{\mathrm{I}} = 0, C_{34}^{\mathrm{I}} = \frac{B_1^0}{12}, \tag{A13}
$$

$$
B_{13}^{\mathrm{I}} = 0, B_{14}^{\mathrm{I}} = -\frac{1}{48}C_1^0 - \frac{1}{2}C_3^0, B_{15}^{\mathrm{I}} = 0, B_{16}^{\mathrm{I}} = -\frac{1}{24}C_5^0 - C_7^0, B_{17}^{\mathrm{I}} = \frac{C_2^0}{8\bar{\lambda}_{33}}. \tag{A14}
$$

(3) The unknown functions $g_i^{\mathrm{II}}(Z)$ and $f_i^{\mathrm{II}}(Z)$ (i = 1,2,3,…,18):

From Equations (43)–(48) and (54), we may obtain the unknown functions of $g_i^{\mathrm{II}}(Z)$ and $f_i^{\mathrm{II}}(Z)$ $(i = 1,2,3,\ldots,18)$,

$$
\begin{cases}
g_i^{\mathrm{II}}(Z) = K_i Z^2 + B_{2i-1}^{\mathrm{II}}Z + B_{2i}^{\mathrm{II}}(i = 1, 2, 10, 11, 13, 14) \\
g_i^{\mathrm{II}}(Z) = B_{2i-1}^{\mathrm{II}}Z + B_{2i}^{\mathrm{II}}(i = 4, 5, 7, 8, 16, 17)
\end{cases} \tag{A15}
$$

$$
\begin{cases}
g_3^{\mathrm{II}}(Z) = -\frac{C_1^{\mathrm{I}}}{24(\bar{\lambda}_{33})^2}Z^4 - \frac{1}{3\bar{\lambda}_{33}}B_1^{\mathrm{II}}Z^3 - \frac{1}{\bar{\lambda}_{33}}B_2^{\mathrm{II}}Z^2 - \frac{(2S_{13}+S_{44})}{24\bar{\lambda}_{33}}C_1^{\mathrm{I}}Z^4 \\
\quad - \frac{(2S_{13}+S_{44})}{6\bar{\lambda}_{33}}C_2^{\mathrm{I}}Z^3 - \frac{1}{3(\bar{\lambda}_{33})^2}B_1^0 Z^3 + \frac{1}{2\bar{\lambda}_{33}}C_9^{\mathrm{I}}Z^2 + B_5^{\mathrm{II}}Z + B_6^{\mathrm{II}} \\
g_6^{\mathrm{II}}(Z) = -\frac{1}{3\bar{\lambda}_{33}}B_7^{\mathrm{II}}Z^3 - \frac{1}{\bar{\lambda}_{33}}B_8^{\mathrm{II}}Z^2 + \frac{1}{24\bar{\lambda}_{33}}C_{13}^{\mathrm{I}}Z^4 + \frac{1}{6\bar{\lambda}_{33}}C_{14}^{\mathrm{I}}Z^3 + \frac{1}{2\bar{\lambda}_{33}}C_{15}^{\mathrm{I}}Z^2 + B_{11}^{\mathrm{II}}Z + B_{12}^{\mathrm{II}} \\
g_9^{\mathrm{II}}(Z) = -\frac{1}{3\bar{\lambda}_{33}}B_{13}^{\mathrm{II}}Z^3 - \frac{1}{\bar{\lambda}_{33}}B_{14}^{\mathrm{II}}Z^2 - \frac{1}{24\bar{\lambda}_{33}}C_{25}^{\mathrm{I}}Z^4 - \frac{C_{26}^{\mathrm{I}}}{6\bar{\lambda}_{33}}Z^3 - \frac{C_{27}^{\mathrm{I}}}{2\bar{\lambda}_{33}}Z^2 + B_{17}^{\mathrm{II}}Z + B_{18}^{\mathrm{II}} \\
g_{12}^{\mathrm{II}}(Z) = -\frac{C_{13}^{\mathrm{I}}}{24(\bar{\lambda}_{33})^2}Z^4 - \frac{1}{3\bar{\lambda}_{33}}B_{19}^{\mathrm{II}}Z^3 - \frac{1}{\bar{\lambda}_{33}}B_{20}^{\mathrm{II}}Z^2 + \frac{1}{24\bar{\lambda}_{33}}C_1^{\mathrm{I}}Z^4 + \frac{1}{6\bar{\lambda}_{33}}C_2^{\mathrm{I}}Z^3 + \frac{1}{2\bar{\lambda}_{33}}C_3^{\mathrm{I}}Z^2 \\
\quad - \frac{(2S_{13}+S_{44})}{24\bar{\lambda}_{33}}C_{13}^{\mathrm{I}}Z^4 - \frac{(2S_{13}+S_{44})}{6\bar{\lambda}_{33}}C_{14}^{\mathrm{I}}Z^3 + \frac{1}{3\bar{\lambda}_{33}}B_1^0 Z^3 + \frac{1}{2\bar{\lambda}_{33}}C_{21}^{\mathrm{I}}Z^2 + B_{23}^{\mathrm{II}}Z + B_{24}^{\mathrm{II}} \\
g_{15}^{\mathrm{II}}(Z) = -\frac{C_{25}^{\mathrm{I}}}{24(\bar{\lambda}_{33})^2}Z^4 - \frac{1}{3\bar{\lambda}_{33}}B_{25}^{\mathrm{II}}Z^3 - \frac{1}{\bar{\lambda}_{33}}B_{26}^{\mathrm{II}}Z^2 - \frac{(2S_{13}+S_{44})}{24\bar{\lambda}_{33}}C_{25}^{\mathrm{I}}Z^4 - \frac{(2S_{13}+S_{44})}{6\bar{\lambda}_{33}}C_{26}^{\mathrm{I}}Z^3 \\
\quad - \frac{1}{3\bar{\lambda}_{33}}B_1^0 Z^3 + \frac{C_{33}^{\mathrm{I}}}{2\bar{\lambda}_{33}}Z^2 - \frac{1}{24\bar{\lambda}_{33}}C_1^{\mathrm{I}}Z^4 - \frac{C_2^{\mathrm{I}}}{6\bar{\lambda}_{33}}Z^3 - \frac{C_3^{\mathrm{I}}}{2\bar{\lambda}_{33}}Z^2 + B_{29}^{\mathrm{II}}Z + B_{30}^{\mathrm{II}} \\
g_{18}^{\mathrm{II}}(Z) = -\frac{1}{3\bar{\lambda}_{33}}B_{31}^{\mathrm{II}}Z^3 - \frac{1}{\bar{\lambda}_{33}}B_{32}^{\mathrm{II}}Z^2 + \frac{1}{24\bar{\lambda}_{33}}C_{25}^{\mathrm{I}}Z^4 + \frac{C_{26}^{\mathrm{I}}}{6\bar{\lambda}_{33}}Z^3 + \frac{C_{27}^{\mathrm{I}}}{2\bar{\lambda}_{33}}Z^2 \\
\quad - \frac{C_{13}^{\mathrm{I}}}{24\bar{\lambda}_{33}}Z^4 - \frac{C_{14}^{\mathrm{I}}}{6\bar{\lambda}_{33}}Z^3 - \frac{C_{15}^{\mathrm{I}}}{2\bar{\lambda}_{33}}Z^2 + B_{35}^{\mathrm{II}}Z + B_{36}^{\mathrm{II}}
\end{cases} \tag{A16}
$$

$$f_i^{II}(Z) = \frac{1}{6}C_{4i-3}^{II}Z^3 + \frac{1}{2}C_{4i-2}^{II}Z^2 + C_{4i-1}^{II}Z$$
$$+ C_{4i}^{II} (i = 1,2,4,5,7,8,10,11,13,14,16,17) \quad , \tag{A17}$$

$$\begin{cases}
f_3^{II}(Z) = -\frac{(2S_{13}+S_{44})}{120}C_1^{II}Z^5 - \frac{(2S_{13}+S_{44})}{24}C_2^{II}Z^4 - \frac{C_1^0}{120(\overline{\lambda}_{33})^2}Z^5 - \frac{1}{12\overline{\lambda}_{33}}B_1^I Z^4 \\
\quad -\frac{(2S_{13}+S_{44})}{120\overline{\lambda}_{33}}C_1^0 Z^5 - \frac{(2S_{13}+S_{44})}{24\overline{\lambda}_{33}}C_2^0 Z^4 + \frac{1}{6}C_9^{II}Z^3 + \frac{1}{2}C_{10}^{II}Z^2 + C_{11}^{II}Z + C_{12}^{II} \\
f_6^{II}(Z) = -\frac{(2S_{13}+S_{44})}{120}C_{13}^{II}Z^5 - \frac{(2S_{13}+S_{44})}{24}C_{14}^{II}Z^4 + \frac{1}{12}B_7^I Z^4 \\
\quad +\frac{1}{6}C_{21}^{II}Z^3 + \frac{1}{2}C_{22}^{II}Z^2 + C_{23}^{II}Z + C_{24}^{II} \\
f_9^{II}(Z) = -\frac{(2S_{13}+S_{44})}{120}C_{25}^{II}Z^5 - \frac{(2S_{13}+S_{44})}{24}C_{26}^{II}Z^4 - \frac{1}{12}B_{13}^I Z^4 \\
\quad +\frac{1}{6}C_{33}^{II}Z^3 + \frac{1}{2}C_{34}^{II}Z^2 + C_{35}^{II}Z + C_{36}^{II} \\
f_{12}^{II}(Z) = -\frac{(2S_{13}+S_{44})}{120}C_{37}^{II}Z^5 - \frac{(2S_{13}+S_{44})}{24}C_{38}^{II}Z^4 - \frac{1}{12\overline{\lambda}_{33}}B_7^I Z^4 + \frac{C_1^0}{60\overline{\lambda}_{33}}Z^5 \\
\quad +\frac{C_2^0}{24\overline{\lambda}_{33}}Z^4 + \frac{1}{12}B_1^I Z^4 + \frac{1}{6}C_{45}^{II}Z^3 + \frac{1}{2}C_{46}^{II}Z^2 + C_{47}^{II}Z + C_{48}^{II} \\
f_{15}^{II}(Z) = -\frac{(2S_{13}+S_{44})}{120}C_{49}^{II}Z^5 - \frac{(2S_{13}+S_{44})}{24}C_{50}^{II}Z^4 - \frac{1}{12\overline{\lambda}_{33}}B_{13}^I Z^4 - \frac{C_1^0}{60\overline{\lambda}_{33}}Z^5 \\
\quad -\frac{C_2^0}{24\overline{\lambda}_{33}}Z^4 - \frac{1}{12}B_1^I Z^4 + \frac{1}{6}C_{57}^{II}Z^3 + \frac{1}{2}C_{58}^{II}Z^2 + C_{59}^{II}Z + C_{60}^{II} \\
f_{18}^{II}(Z) = -\frac{(2S_{13}+S_{44})}{120}C_{61}^{II}Z^5 - \frac{(2S_{13}+S_{44})}{24}C_{62}^{II}Z^4 + \frac{1}{12}B_{13}^I Z^4 \\
\quad -\frac{1}{12}B_7^I Z^4 + \frac{1}{6}C_{69}^{II}Z^3 + \frac{1}{2}C_{70}^{II}Z^2 + C_{71}^{II}Z + C_{72}^{II}
\end{cases} \tag{A18}$$

and

$$K_1 = \frac{C_1^I}{4\overline{\lambda}_{33}}, K_2 = \frac{C_5^I}{2\overline{\lambda}_{33}}, K_{10} = \frac{C_{13}^I}{4\overline{\lambda}_{33}}, K_{11} = \frac{C_{17}^I}{2\overline{\lambda}_{33}}, K_{13} = \frac{C_{25}^I}{4\overline{\lambda}_{33}}, K_{14} = \frac{C_{29}^I}{2\overline{\lambda}_{33}}, \tag{A19}$$

where B_i^{II} ($i = 1,2,3,\ldots,36$) and C_i^{II} ($i = 1,2,3,\ldots,72$) are undetermined constants which can be determined by Equations (49)–(53),

$$C_1^{II} = 0, C_2^{II} = 0, C_3^{II} = 0, C_4^{II} = 0, C_5^{II} = 0, C_6^{II} = 0, C_7^{II} = 0,$$
$$C_9^{II} = \frac{1}{40(\overline{\lambda}_{33})^2}C_1^0 + \frac{(2S_{13}+S_{44})}{40\overline{\lambda}_{33}}C_1^0, C_{10}^{II} = \frac{1}{12\overline{\lambda}_{33}}B_1^I + \frac{(2S_{13}+S_{44})}{24\overline{\lambda}_{33}}C_2^0 , \tag{A20}$$

$$B_1^{II} = \frac{C_2^I}{2\overline{\lambda}_{33}}, B_2^{II} = -\frac{C_1^I}{48\overline{\lambda}_{33}}, B_3^{II} = \frac{C_6^I}{\overline{\lambda}_{33}}, B_4^{II} = -\frac{C_5^I}{24\overline{\lambda}_{33}}, B_5^{II} = \frac{C_{10}^I}{\overline{\lambda}_{33}} + \frac{C_2^I}{8(\overline{\lambda}_{33})^2}, \tag{A21}$$

$$C_{13}^{II} = 0, C_{14}^{II} = 0, C_{15}^{II} = 0, C_{16}^{II} = 0, C_{17}^{II} = 0, C_{18}^{II} = 0, C_{19}^{II} = 0, C_{21}^{II} = 0, C_{22}^{II} = -\frac{1}{12}B_7^I, \tag{A22}$$

$$B_7^{II} = 0, B_8^{II} = 0, B_9^{II} = 0, B_{10}^{II} = 0, B_{11}^{II} = \frac{C_{16}^I}{\overline{\lambda}_{33}}, \tag{A23}$$

$$C_{25}^{II} = 0, C_{26}^{II} = 0, C_{27}^{II} = 0, C_{28}^{II} = 0, C_{29}^{II} = 0, C_{30}^{II} = 0, C_{31}^{II} = 0, C_{33}^{II} = 0, C_{34}^{II} = \frac{1}{12}B_{13}^I, \tag{A24}$$

$$B_{13}^{II} = 0, B_{14}^{II} = -\frac{1}{48}C_{25}^I - \frac{1}{2}C_{27}^I, B_{15}^{II} = 0, B_{16}^{II} = -\frac{1}{24}C_{29}^I - C_{31}^I, B_{17}^{II} = \frac{C_{26}^I}{8\overline{\lambda}_{33}}, \tag{A25}$$

$$C_{37}^{II} = 0, C_{38}^{II} = 0, C_{39}^{II} = 0, C_{40}^{II} = 0, C_{41}^{II} = 0, C_{42}^{II} = 0, C_{43}^{II} = 0,$$
$$C_{45}^{II} = -\frac{C_1^0}{20\overline{\lambda}_{33}}, C_{46}^{II} = \frac{1}{12\overline{\lambda}_{33}}B_7^I - \frac{C_2^0}{24\overline{\lambda}_{33}} - \frac{1}{12}B_1^I , \tag{A26}$$

$$B_{19}^{II} = \frac{C_{14}^I}{2\overline{\lambda}_{33}}, B_{20}^{II} = -\frac{C_{13}^I}{48\overline{\lambda}_{33}}, B_{21}^{II} = \frac{C_{18}^I}{\overline{\lambda}_{33}}, B_{22}^{II} = -\frac{C_{17}^I}{24\overline{\lambda}_{33}}, B_{23}^{II} = \frac{C_{22}^I}{\overline{\lambda}_{33}} + \frac{C_4^I}{\overline{\lambda}_{33}} + \frac{C_{14}^I}{8(\overline{\lambda}_{33})^2}, \tag{A27}$$

$$C_{49}^{II} = 0, C_{50}^{II} = 0, C_{51}^{II} = 0, C_{52}^{II} = 0, C_{53}^{II} = 0, C_{54}^{II} = 0, C_{55}^{II} = 0,$$
$$C_{57}^{II} = \frac{C_1^0}{20\bar{\lambda}_{33}}, C_{58}^{II} = \frac{1}{12\bar{\lambda}_{33}}B_{13}^I + \frac{C_2^0}{24\bar{\lambda}_{33}} + \frac{1}{12}B_1^I \tag{A28}$$

$$B_{25}^{II} = \frac{C_{26}^I}{2\bar{\lambda}_{33}}, B_{26}^{II} = -\frac{C_1^I}{48} - \frac{C_3^I}{2} - \frac{C_{25}^I}{48\bar{\lambda}_{33}}, B_{27}^{II} = \frac{C_{30}^I}{\bar{\lambda}_{33}}$$
$$B_{28}^{II} = -\frac{1}{24}C_5^I - C_7^I - \frac{C_{29}^I}{24\bar{\lambda}_{33}}, B_{29}^{II} = \frac{C_{34}^I}{\bar{\lambda}_{33}} + \frac{C_{26}^I}{8(\bar{\lambda}_{33})^2} + \frac{C_2^I}{8\bar{\lambda}_{33}} \tag{A29}$$

$$C_{61}^{II} = 0, C_{62}^{II} = 0, C_{63}^{II} = 0, C_{64}^{II} = 0, C_{65}^{II} = 0, C_{66}^{II} = 0, C_{67}^{II} = 0,$$
$$C_{69}^{II} = 0, C_{70}^{II} = -\frac{1}{12}B_{13}^I + \frac{1}{12}B_7^I \tag{A30}$$

and

$$B_{31}^{II} = 0, B_{32}^{II} = -\frac{1}{48}C_{13}^I - \frac{1}{2}C_{15}^I, B_{33}^{II} = 0,$$
$$B_{34}^{II} = -\frac{1}{24}C_{17}^I - C_{19}^I, B_{35}^{II} = \frac{C_{28}^I}{\bar{\lambda}_{33}} + \frac{C_{14}^I}{8\bar{\lambda}_{33}} \tag{A31}$$

Appendix B

From Equations (3)–(4), (16), and (55)–(56), we have

$$\frac{\partial \bar{w}}{\partial Z} = S_{13}\frac{\partial^2 \bar{U}}{\partial Z^2} + S_{33}\frac{\partial^2 U}{\partial X^2} - \bar{d}_{33}\frac{\partial \Phi}{\partial Z}$$
$$= -6S_{13}\bar{q}X^2Z - \frac{12}{b}S_{13}\bar{P}XZ + 2(2S_{13} + S_{44})S_{13}\bar{q}Z^3 - \frac{3(2S_{13}+S_{44})}{10}S_{13}\bar{q}Z$$
$$+ \frac{12}{b}S_{13}\bar{M}Z + S_{13}(\bar{d}_{31})^2[\frac{2}{(\bar{\lambda}_{33})^2}\bar{q}Z^3 + \frac{2(2S_{13}+S_{44})}{\bar{\lambda}_{33}}\bar{q}Z^3 - \frac{3}{10(\bar{\lambda}_{33})^2}\bar{q}Z - \frac{3(2S_{13}+S_{44})}{10\bar{\lambda}_{33}}\bar{q}Z]$$
$$+ S_{13}\bar{d}_{31}\bar{d}_{33}(-\frac{4}{\bar{\lambda}_{33}}\bar{q}Z^3 + \frac{3}{5\bar{\lambda}_{33}}\bar{q}Z) + S_{13}\bar{d}_{31}\bar{d}_{15}(\frac{4}{\bar{\lambda}_{33}}\bar{q}Z^3 - \frac{3}{5\bar{\lambda}_{33}}\bar{q}Z) - 2S_{33}\bar{q}Z^3 + \frac{3}{2}S_{33}\bar{q}Z - S_{33}\frac{\bar{q}}{2}$$
$$- \bar{d}_{33}\bar{d}_{31}[-\frac{6}{\bar{\lambda}_{33}}\bar{q}X^2Z - \frac{12}{b\bar{\lambda}_{33}}\bar{P}XZ + \frac{2}{(\bar{\lambda}_{33})^2}\bar{q}Z^3 - \frac{1}{2(\bar{\lambda}_{33})^2}\bar{q}Z + \frac{2(2S_{13}+S_{44})}{\bar{\lambda}_{33}}\bar{q}Z^3 + \frac{12}{b\bar{\lambda}_{33}}\bar{M}Z$$
$$- \frac{3(2S_{13}+S_{44})}{10\bar{\lambda}_{33}}\bar{q}Z] - \bar{d}_{33}\bar{d}_{33}[-\frac{2}{\bar{\lambda}_{33}}\bar{q}Z^3 + \frac{3}{2\bar{\lambda}_{33}}\bar{q}Z - \frac{1}{2\bar{\lambda}_{33}}\bar{q}] - \bar{d}_{33}\bar{d}_{15}[\frac{1}{\bar{\lambda}_{33}}\bar{q}Z + \frac{2}{\bar{\lambda}_{33}}\bar{q}Z^3 - \frac{3}{2\bar{\lambda}_{33}}\bar{q}Z] \tag{A32}$$

and

$$\frac{\partial \bar{u}}{\partial X} = \frac{\partial^2 \bar{U}}{\partial Z^2} + S_{13}\frac{\partial^2 \bar{U}}{\partial X^2} - \bar{d}_{31}\frac{\partial \Phi}{\partial Z}$$
$$= -6\bar{q}X^2Z - \frac{12}{b}\bar{P}XZ + 2(2S_{13} + S_{44})\bar{q}Z^3 + \frac{12}{b}\bar{M}Z - \frac{3(2S_{13}+S_{44})}{10}\bar{q}Z + (\bar{d}_{31})^2[\frac{2}{(\bar{\lambda}_{33})^2}\bar{q}Z^3$$
$$+ \frac{2(2S_{13}+S_{44})}{\bar{\lambda}_{33}}\bar{q}Z^3 - \frac{3}{10(\bar{\lambda}_{33})^2}\bar{q}Z - \frac{3(2S_{13}+S_{44})}{10\bar{\lambda}_{33}}\bar{q}Z] + \bar{d}_{31}\bar{d}_{33}(-\frac{4}{\bar{\lambda}_{33}}\bar{q}Z^3 + \frac{3}{5\bar{\lambda}_{33}}\bar{q}Z)$$
$$+ \bar{d}_{31}\bar{d}_{15}(\frac{4}{\bar{\lambda}_{33}}\bar{q}Z^3 - \frac{3}{5\bar{\lambda}_{33}}\bar{q}Z) - 2S_{13}\bar{q}Z^3 + \frac{3}{2}S_{13}\bar{q}Z - S_{13}\frac{\bar{q}}{2} - \bar{d}_{31}\bar{d}_{31}[-\frac{6}{\bar{\lambda}_{33}}\bar{q}X^2Z$$
$$- \frac{12}{b\bar{\lambda}_{33}}\bar{P}XZ + \frac{2}{(\bar{\lambda}_{33})^2}\bar{q}Z^3 - \frac{1}{2(\bar{\lambda}_{33})^2}\bar{q}Z + \frac{2(2S_{13}+S_{44})}{\bar{\lambda}_{33}}\bar{q}Z^3 + \frac{12}{b\bar{\lambda}_{33}}\bar{M}Z - \frac{3(2S_{13}+S_{44})}{10\bar{\lambda}_{33}}\bar{q}Z]$$
$$- \bar{d}_{31}\bar{d}_{33}[-\frac{2}{\bar{\lambda}_{33}}\bar{q}Z^3 + \frac{3}{2\bar{\lambda}_{33}}\bar{q}Z - \frac{1}{2\bar{\lambda}_{33}}\bar{q}] - \bar{d}_{31}\bar{d}_{15}[\frac{1}{\bar{\lambda}_{33}}\bar{q}Z + \frac{2}{\bar{\lambda}_{33}}\bar{q}Z^3 - \frac{3}{2\bar{\lambda}_{33}}\bar{q}Z] \tag{A33}$$

Integrating Equations (A32) and (A33) respectively, we may obtain

$$\bar{w} = -\frac{6}{b}S_{13}\bar{P}XZ^2 + \frac{(2S_{13}+S_{44})}{2}S_{13}\bar{q}Z^4 + \frac{6}{b}S_{13}\bar{M}Z^2 - \frac{3(2S_{13}+S_{44})}{20}S_{13}\bar{q}Z^2 - 3S_{13}\bar{q}X^2Z^2$$
$$+ S_{13}(\bar{d}_{31})^2[\frac{1}{2(\bar{\lambda}_{33})^2}\bar{q}Z^4 + \frac{(2S_{13}+S_{44})}{2\bar{\lambda}_{33}}\bar{q}Z^4 - \frac{3}{20(\bar{\lambda}_{33})^2}\bar{q}Z^2 - \frac{3(2S_{13}+S_{44})}{20\bar{\lambda}_{33}}\bar{q}Z^2]$$
$$+ S_{13}\bar{d}_{31}\bar{d}_{33}(-\frac{1}{\bar{\lambda}_{33}}\bar{q}Z^4 + \frac{3}{10\bar{\lambda}_{33}}\bar{q}Z^2) + S_{13}\bar{d}_{31}\bar{d}_{15}(\frac{1}{\bar{\lambda}_{33}}\bar{q}Z^4 - \frac{3}{10\bar{\lambda}_{33}}\bar{q}Z^2) - \frac{S_{33}}{2}\bar{q}Z^4$$
$$+ \frac{3}{4}S_{33}\bar{q}Z^2 - S_{33}\frac{\bar{q}}{2}Z - \bar{d}_{33}\bar{d}_{31}[-\frac{3}{\bar{\lambda}_{33}}\bar{q}X^2Z^2 - \frac{6}{b\bar{\lambda}_{33}}\bar{P}XZ^2 + \frac{1}{2(\bar{\lambda}_{33})^2}\bar{q}Z^4 - \frac{1}{4(\bar{\lambda}_{33})^2}\bar{q}Z^2$$
$$+ \frac{(2S_{13}+S_{44})}{2\bar{\lambda}_{33}}\bar{q}Z^4 + \frac{6}{b\bar{\lambda}_{33}}\bar{M}Z^2 - \frac{3(2S_{13}+S_{44})}{20\bar{\lambda}_{33}}\bar{q}Z^2] - \bar{d}_{33}\bar{d}_{33}[-\frac{1}{2\bar{\lambda}_{33}}\bar{q}Z^4$$
$$+ \frac{3}{4\bar{\lambda}_{33}}\bar{q}Z^2 - \frac{1}{2\bar{\lambda}_{33}}\bar{q}Z] - \bar{d}_{33}\bar{d}_{15}[\frac{1}{2\bar{\lambda}_{33}}\bar{q}Z^2 + \frac{1}{2\bar{\lambda}_{33}}\bar{q}Z^4 - \frac{3}{4\bar{\lambda}_{33}}\bar{q}Z^2] + G_0(X) \tag{A34}$$

and

$$\overline{u} = -2\overline{q}X^3Z - \tfrac{6}{b}\overline{P}X^2Z + 2(2S_{13}+S_{44})\overline{q}XZ^3 + \tfrac{12}{b}\overline{M}XZ - \tfrac{3(2S_{13}+S_{44})}{10}\overline{q}XZ$$

$$+(\overline{d}_{31})^2[\tfrac{2}{(\overline{\lambda}_{33})^2}\overline{q}XZ^3 + \tfrac{2(2S_{13}+S_{44})}{\overline{\lambda}_{33}}\overline{q}XZ^3 - \tfrac{3}{10(\overline{\lambda}_{33})^2}\overline{q}XZ - \tfrac{3(2S_{13}+S_{44})}{10\overline{\lambda}_{33}}\overline{q}XZ]$$

$$+\overline{d}_{31}\overline{d}_{33}(-\tfrac{4}{\overline{\lambda}_{33}}\overline{q}XZ^3 + \tfrac{3}{5\overline{\lambda}_{33}}\overline{q}XZ) + \overline{d}_{31}\overline{d}_{15}(\tfrac{4}{\overline{\lambda}_{33}}\overline{q}XZ^3 - \tfrac{3}{5\overline{\lambda}_{33}}\overline{q}XZ) - 2S_{13}\overline{q}XZ^3$$

$$+\tfrac{3}{2}S_{13}\overline{q}XZ - S_{13}\tfrac{\overline{q}}{2}X - \overline{d}_{31}\overline{d}_{31}[-\tfrac{2}{\overline{\lambda}_{33}}\overline{q}X^3Z - \tfrac{6}{b\overline{\lambda}_{33}}\overline{P}X^2Z + \tfrac{2}{(\overline{\lambda}_{33})^2}\overline{q}XZ^3 - \tfrac{1}{2(\overline{\lambda}_{33})^2}\overline{q}XZ \qquad ,$$

$$+\tfrac{2(2S_{13}+S_{44})}{\overline{\lambda}_{33}}\overline{q}XZ^3 + \tfrac{12}{b\overline{\lambda}_{33}}\overline{M}XZ - \tfrac{3(2S_{13}+S_{44})}{10\overline{\lambda}_{33}}\overline{q}XZ] - \overline{d}_{31}\overline{d}_{33}[-\tfrac{2}{\overline{\lambda}_{33}}\overline{q}XZ^3$$

$$+\tfrac{3}{2\overline{\lambda}_{33}}\overline{q}XZ - \tfrac{1}{2\overline{\lambda}_{33}}\overline{q}X] - \overline{d}_{31}\overline{d}_{15}[\tfrac{1}{\overline{\lambda}_{33}}\overline{q}XZ + \tfrac{2}{\overline{\lambda}_{33}}\overline{q}XZ^3 - \tfrac{3}{2\overline{\lambda}_{33}}\overline{q}XZ] + G_1(Z) \tag{A35}$$

From Equations (5) and (16), one has

$$-2\overline{q}X^3 + \tfrac{2}{\overline{\lambda}_{33}}\overline{d}_{31}\overline{d}_{31}\overline{q}X^3 - \tfrac{6}{b}\overline{P}X^2 + \tfrac{6}{b\overline{\lambda}_{33}}\overline{d}_{31}\overline{d}_{31}\overline{P}X^2 + \tfrac{12}{b}\overline{M}X + \tfrac{9}{10}S_{13}\overline{q}X + \tfrac{12}{10}S_{44}\overline{q}X$$

$$+\tfrac{1}{5(\overline{\lambda}_{33})^2}\overline{d}_{31}\overline{d}_{31}\overline{q}X - \tfrac{9}{10\overline{\lambda}_{33}}\overline{d}_{31}\overline{d}_{33}\overline{q}X + \tfrac{2}{5\overline{\lambda}_{33}}\overline{d}_{31}\overline{d}_{15}\overline{q}X - \tfrac{12}{b\overline{\lambda}_{33}}\overline{d}_{31}\overline{d}_{31}\overline{M}X - \overline{d}_{15}\overline{d}_{15}\overline{q}X + \tfrac{dG_0(X)}{dX} \qquad , \tag{A36}$$

$$= [\tfrac{6}{b}S_{13}\overline{P} + \tfrac{6}{b}S_{44}\overline{P} - \tfrac{6}{b\overline{\lambda}_{33}}\overline{d}_{33}\overline{d}_{31}\overline{P} + \tfrac{6}{b\overline{\lambda}_{33}}\overline{d}_{15}\overline{d}_{31}\overline{P}]Z^2 - \tfrac{3}{2b}S_{44}\overline{P} - \tfrac{3}{2b\overline{\lambda}_{33}}\overline{d}_{15}\overline{d}_{31}\overline{P} - \tfrac{dG_1(Z)}{dZ} = A$$

where A is an undetermined constant. From Equation (A36), we have

$$\begin{cases} \tfrac{dG_0(X)}{dX} = [2\overline{q} - \tfrac{2}{\overline{\lambda}_{33}}\overline{d}_{31}\overline{d}_{31}\overline{q}]X^3 + [\tfrac{6}{b}\overline{P} - \tfrac{6}{b\overline{\lambda}_{33}}\overline{d}_{31}\overline{d}_{31}\overline{P}]X^2 + [-\tfrac{12}{b}\overline{M} - \tfrac{9}{10}S_{13}\overline{q} - \tfrac{12}{10}S_{44}\overline{q} \\[2mm] -\tfrac{1}{5(\overline{\lambda}_{33})^2}\overline{d}_{31}\overline{d}_{31}\overline{q} + \tfrac{9}{10\overline{\lambda}_{33}}\overline{d}_{31}\overline{d}_{33}\overline{q} - \tfrac{2}{5\overline{\lambda}_{33}}\overline{d}_{31}\overline{d}_{15}\overline{q} + \tfrac{12}{b\overline{\lambda}_{33}}\overline{d}_{31}\overline{d}_{31}\overline{M} + \overline{d}_{15}\overline{d}_{15}\overline{q}]X + A \\[2mm] \tfrac{dG_1(Z)}{dZ} = [\tfrac{6}{b}S_{13}\overline{P} + \tfrac{6}{b}S_{44}\overline{P} - \tfrac{6}{b\overline{\lambda}_{33}}\overline{d}_{33}\overline{d}_{31}\overline{P} + \tfrac{6}{b\overline{\lambda}_{33}}\overline{d}_{15}\overline{d}_{31}\overline{P}]Z^2 - \tfrac{3}{2b}S_{44}\overline{P} - \tfrac{3}{2b\overline{\lambda}_{33}}\overline{d}_{15}\overline{d}_{31}\overline{P} - A \end{cases} \tag{A37}$$

Integrating Equation (A37), it can be obtained

$$\begin{cases} G_0(X) = [\tfrac{1}{2}\overline{q} - \tfrac{1}{2\overline{\lambda}_{33}}\overline{d}_{31}\overline{d}_{31}\overline{q}]X^4 + [\tfrac{2}{b}\overline{P} - \tfrac{2}{b\overline{\lambda}_{33}}\overline{d}_{31}\overline{d}_{31}\overline{P}]X^3 + [-\tfrac{6}{b}\overline{M} - \tfrac{9}{20}S_{13}\overline{q} - \tfrac{6}{10}S_{44}\overline{q} \\[2mm] -\tfrac{1}{10(\overline{\lambda}_{33})^2}\overline{d}_{31}\overline{d}_{31}\overline{q} + \tfrac{9}{20\overline{\lambda}_{33}}\overline{d}_{31}\overline{d}_{33}\overline{q} - \tfrac{1}{5\overline{\lambda}_{33}}\overline{d}_{31}\overline{d}_{15}\overline{q} + \tfrac{6}{b\overline{\lambda}_{33}}\overline{d}_{31}\overline{d}_{31}\overline{M} + \tfrac{1}{2}\overline{d}_{15}\overline{d}_{15}\overline{q}]X^2 + AX + B \\[2mm] G_1(Z) = [\tfrac{2}{b}S_{13}\overline{P} + \tfrac{2}{b}S_{44}\overline{P} - \tfrac{2}{b\overline{\lambda}_{33}}\overline{d}_{33}\overline{d}_{31}\overline{P} + \tfrac{2}{b\overline{\lambda}_{33}}\overline{d}_{15}\overline{d}_{31}\overline{P}]Z^3 \\[2mm] +[-\tfrac{3}{2b}S_{44}\overline{P} - \tfrac{3}{2b\overline{\lambda}_{33}}\overline{d}_{15}\overline{d}_{31}\overline{P} - A]Z + C \end{cases} \tag{A38}$$

Thus, we can finally obtain

$$\overline{w} = -3S_{13}\overline{q}X^2Z^2 - \tfrac{6}{b}S_{13}\overline{P}XZ^2 + \tfrac{(2S_{13}+S_{44})}{2}S_{13}\overline{q}Z^4 + \tfrac{6}{b}S_{13}\overline{M}Z^2 - \tfrac{3(2S_{13}+S_{44})}{20}S_{13}\overline{q}Z^2$$

$$+S_{13}(\overline{d}_{31})^2[\tfrac{1}{2(\overline{\lambda}_{33})^2}\overline{q}Z^4 + \tfrac{(2S_{13}+S_{44})}{2\overline{\lambda}_{33}}\overline{q}Z^4 - \tfrac{3}{20(\overline{\lambda}_{33})^2}\overline{q}Z^2 - \tfrac{3(2S_{13}+S_{44})}{20\overline{\lambda}_{33}}\overline{q}Z^2] - \tfrac{S_{33}}{2}\overline{q}Z^4$$

$$+S_{13}\overline{d}_{31}\overline{d}_{33}(-\tfrac{1}{\overline{\lambda}_{33}}\overline{q}Z^4 + \tfrac{3}{10\overline{\lambda}_{33}}\overline{q}Z^2) + S_{13}\overline{d}_{31}\overline{d}_{15}(\tfrac{1}{\overline{\lambda}_{33}}\overline{q}Z^4 - \tfrac{3}{10\overline{\lambda}_{33}}\overline{q}Z^2) + \tfrac{3}{4}S_{33}\overline{q}Z^2 - S_{33}\tfrac{\overline{q}}{2}Z$$

$$-\overline{d}_{33}\overline{d}_{31}[-\tfrac{3}{\overline{\lambda}_{33}}\overline{q}X^2Z^2 - \tfrac{6}{b\overline{\lambda}_{33}}\overline{P}XZ^2 + \tfrac{1}{2(\overline{\lambda}_{33})^2}\overline{q}Z^4 - \tfrac{1}{4(\overline{\lambda}_{33})^2}\overline{q}Z^2 + \tfrac{(2S_{13}+S_{44})}{2\overline{\lambda}_{33}}\overline{q}Z^4 + \tfrac{6}{b\overline{\lambda}_{33}}\overline{M}Z^2 \tag{A39}$$

$$-\tfrac{3(2S_{13}+S_{44})}{20\overline{\lambda}_{33}}\overline{q}Z^2] - \overline{d}_{33}\overline{d}_{33}[-\tfrac{1}{2\overline{\lambda}_{33}}\overline{q}Z^4 + \tfrac{3}{4\overline{\lambda}_{33}}\overline{q}Z^2 - \tfrac{1}{2\overline{\lambda}_{33}}\overline{q}Z] - \overline{d}_{33}\overline{d}_{15}[\tfrac{1}{2\overline{\lambda}_{33}}\overline{q}Z^2 + \tfrac{1}{2\overline{\lambda}_{33}}\overline{q}Z^4$$

$$-\tfrac{3}{4\overline{\lambda}_{33}}\overline{q}Z^2] + [\tfrac{1}{2}\overline{q} - \tfrac{1}{2\overline{\lambda}_{33}}\overline{d}_{31}\overline{d}_{31}\overline{q}]X^4 + [\tfrac{2}{b}\overline{P} - \tfrac{2}{b\overline{\lambda}_{33}}\overline{d}_{31}\overline{d}_{31}\overline{P}]X^3 + [-\tfrac{6}{b}\overline{M} - \tfrac{9}{20}S_{13}\overline{q} - \tfrac{6}{10}S_{44}\overline{q}$$

$$-\tfrac{1}{10(\overline{\lambda}_{33})^2}\overline{d}_{31}\overline{d}_{31}\overline{q} + \tfrac{9}{20\overline{\lambda}_{33}}\overline{d}_{31}\overline{d}_{33}\overline{q} - \tfrac{1}{5\overline{\lambda}_{33}}\overline{d}_{31}\overline{d}_{15}\overline{q} + \tfrac{6}{b\overline{\lambda}_{33}}\overline{d}_{31}\overline{d}_{31}\overline{M} + \tfrac{1}{2}\overline{d}_{15}\overline{d}_{15}\overline{q}]X^2 + AX + B$$

and

$$\bar{u} = -2\bar{q}X^3Z - \frac{6}{b}\bar{P}X^2Z + 2(2S_{13}+S_{44})\bar{q}XZ^3 + \frac{12}{b}\overline{M}XZ - \frac{3(2S_{13}+S_{44})}{10}\bar{q}XZ - 2S_{13}\bar{q}XZ^3$$

$$+(\bar{d}_{31})^2[\frac{2}{(\overline{\lambda}_{33})^2}\bar{q}XZ^3 + \frac{2(2S_{13}+S_{44})}{\overline{\lambda}_{33}}\bar{q}XZ^3 - \frac{3}{10(\overline{\lambda}_{33})^2}\bar{q}XZ - \frac{3(2S_{13}+S_{44})}{10\overline{\lambda}_{33}}\bar{q}XZ] - S_{13}\frac{\bar{q}}{2}X$$

$$+\bar{d}_{31}\bar{d}_{33}(-\frac{4}{\overline{\lambda}_{33}}\bar{q}XZ^3 + \frac{3}{5\overline{\lambda}_{33}}\bar{q}XZ) + \bar{d}_{31}\bar{d}_{15}(\frac{4}{\overline{\lambda}_{33}}\bar{q}XZ^3 - \frac{3}{5\overline{\lambda}_{33}}\bar{q}XZ) + \frac{3}{2}S_{13}\bar{q}XZ$$

$$-\bar{d}_{31}\bar{d}_{31}[-\frac{2}{\overline{\lambda}_{33}}\bar{q}X^3Z - \frac{6}{b\overline{\lambda}_{33}}\bar{P}X^2Z + \frac{2}{(\overline{\lambda}_{33})^2}\bar{q}XZ^3 - \frac{1}{2(\overline{\lambda}_{33})^2}\bar{q}XZ + \frac{2(2S_{13}+S_{44})}{\overline{\lambda}_{33}}\bar{q}XZ^3$$

$$+\frac{12}{b\overline{\lambda}_{33}}\overline{M}XZ - \frac{3(2S_{13}+S_{44})}{10\overline{\lambda}_{33}}\bar{q}XZ] - \bar{d}_{31}\bar{d}_{33}[-\frac{2}{\overline{\lambda}_{33}}\bar{q}XZ^3 + \frac{3}{2\overline{\lambda}_{33}}\bar{q}XZ - \frac{1}{2\overline{\lambda}_{33}}\bar{q}X]$$

$$-\bar{d}_{31}\bar{d}_{15}[\frac{1}{\overline{\lambda}_{33}}\bar{q}XZ + \frac{2}{\overline{\lambda}_{33}}\bar{q}XZ^3 - \frac{3}{2\overline{\lambda}_{33}}\bar{q}XZ] + [\frac{2}{b}S_{13}\bar{P} + \frac{2}{b}S_{44}\bar{P} - \frac{2}{b\overline{\lambda}_{33}}\bar{d}_{33}\bar{d}_{31}\bar{P}$$

$$+\frac{2}{b\overline{\lambda}_{33}}\bar{d}_{15}\bar{d}_{31}\bar{P}]Z^3 + [-\frac{3}{2b}S_{44}\bar{P} - \frac{3}{2b\overline{\lambda}_{33}}\bar{d}_{15}\bar{d}_{31}\bar{P} - A]Z + C \tag{A40}$$

Substituting Equations (A39) and (A40) into Equation (24), we have

$$A = -[2\bar{q} - \frac{2}{\overline{\lambda}_{33}}\bar{d}_{31}\bar{d}_{31}\bar{q}]\frac{l^3}{h^3} - [\frac{6}{b}\bar{P} - \frac{6}{b\overline{\lambda}_{33}}\bar{d}_{31}\bar{d}_{31}\bar{P}]\frac{l^2}{h^2} - [-\frac{12}{b}\overline{M} - \frac{9}{10}S_{13}\bar{q} - \frac{6}{5}S_{44}\bar{q}$$

$$-\frac{1}{5(\overline{\lambda}_{33})^2}\bar{d}_{31}\bar{d}_{31}\bar{q} + \frac{9}{10\overline{\lambda}_{33}}\bar{d}_{31}\bar{d}_{33}\bar{q} - \frac{2}{5\overline{\lambda}_{33}}\bar{d}_{31}\bar{d}_{15}\bar{q} + \frac{12}{b\overline{\lambda}_{33}}\bar{d}_{31}\bar{d}_{31}\overline{M} + \bar{d}_{15}\bar{d}_{15}\bar{q}]\frac{l}{h} \tag{A41}$$

$$B = [\frac{3}{2}\bar{q} - \frac{3}{2\overline{\lambda}_{33}}\bar{d}_{31}\bar{d}_{31}\bar{q}]\frac{l^4}{h^4} + [\frac{4}{b}\bar{P} - \frac{4}{b\overline{\lambda}_{33}}\bar{d}_{31}\bar{d}_{31}\bar{P}]\frac{l^3}{h^3} - [\frac{6}{b}\overline{M} + \frac{9}{20}S_{13}\bar{q} + \frac{6}{10}S_{44}\bar{q}$$

$$+\frac{1}{10(\overline{\lambda}_{33})^2}\bar{d}_{31}\bar{d}_{31}\bar{q} - \frac{9}{20\overline{\lambda}_{33}}\bar{d}_{31}\bar{d}_{33}\bar{q} + \frac{1}{5\overline{\lambda}_{33}}\bar{d}_{31}\bar{d}_{15}\bar{q} - \frac{6}{b\overline{\lambda}_{33}}\bar{d}_{31}\bar{d}_{31}\overline{M} - \frac{1}{2}\bar{d}_{15}\bar{d}_{15}\bar{q}]\frac{l^2}{h^2} \tag{A42}$$

and

$$C = S_{13}\frac{\bar{q}}{2}\frac{l}{h} - \frac{1}{2\overline{\lambda}_{33}}\bar{d}_{31}\bar{d}_{33}\bar{q}\frac{l}{h}. \tag{A43}$$

From Equations (16), (A41), (A42), and (A43), Equations (A39) and (A40) can be transformed into

$$w = -3s_{13}q\frac{1}{h^3}x^2z^2 - \frac{6}{b}s_{13}P\frac{1}{h^3}xz^2 + \frac{(2s_{13}+s_{44})}{2s_{11}}s_{13}q\frac{1}{h^3}z^4 + \frac{6}{b}s_{13}\frac{M}{h^3}z^2 - \frac{3(2s_{13}+s_{44})}{20s_{11}}s_{13}q\frac{1}{h}z^2$$

$$+\frac{\lambda_{11}}{2(\lambda_{33})^2}\frac{s_{13}}{s_{11}}(d_{31})^2q\frac{1}{h^3}z^4 + \frac{(2s_{13}+s_{44})s_{13}}{2\lambda_{33}(s_{11})^2}q(d_{31})^2\frac{1}{h^3}z^4 - \frac{3\lambda_{11}s_{13}}{20(\lambda_{33})^2s_{11}}(d_{31})^2q\frac{1}{h}z^2 - \frac{s_{33}}{2}qz$$

$$-\frac{3(2s_{13}+s_{44})}{20\lambda_{33}(s_{11})^2}s_{13}(d_{31})^2q\frac{1}{h}z^2 - \frac{s_{13}}{\lambda_{33}s_{11}}d_{31}d_{33}q\frac{1}{h^3}z^4 + \frac{3s_{13}}{10\lambda_{33}s_{11}}d_{31}d_{33}q\frac{1}{h}z^2 - \frac{s_{33}}{2}q\frac{1}{h^3}z^4$$

$$+\frac{s_{13}}{\lambda_{33}s_{11}}d_{31}d_{15}q\frac{1}{h^3}z^4 - \frac{3s_{13}}{10\lambda_{33}s_{11}}d_{31}d_{15}q\frac{1}{h}z^2 + \frac{3s_{33}}{4}q\frac{1}{h}z^2 + \frac{3}{\lambda_{33}}d_{33}d_{31}q\frac{1}{h^3}x^2z^2$$

$$+\frac{6}{b\lambda_{33}}d_{33}d_{31}\frac{P}{h^3}xz^2 - \frac{\lambda_{11}}{2(\lambda_{33})^2}d_{33}d_{31}q\frac{1}{h^3}z^4 + \frac{\lambda_{11}}{4(\lambda_{33})^2}d_{33}d_{31}q\frac{1}{h}z^2 - \frac{6}{b\lambda_{33}}d_{33}d_{31}\frac{M}{h^3}z^2$$

$$-\frac{(2s_{13}+s_{44})}{2\lambda_{33}s_{11}}d_{33}d_{31}q\frac{1}{h^3}z^4 + \frac{3(2s_{13}+s_{44})}{20\lambda_{33}s_{11}}d_{33}d_{31}q\frac{1}{h}z^2 + \frac{1}{2\lambda_{33}}d_{33}d_{33}q\frac{1}{h^3}z^4$$

$$-\frac{3}{4\lambda_{33}}d_{33}d_{33}q\frac{1}{h}z^2 + \frac{1}{2\lambda_{33}}d_{33}d_{33}qz - \frac{1}{2\lambda_{33}}d_{33}d_{15}q\frac{1}{h}z^2 - \frac{1}{2\lambda_{33}}d_{33}d_{15}q\frac{1}{h^3}z^4$$

$$+\frac{3}{4\lambda_{33}}d_{33}d_{15}q\frac{1}{h}z^2 + [\frac{1}{2}qs_{11} - \frac{1}{2\lambda_{33}}d_{31}d_{31}q]\frac{1}{h^3}x^4 + [\frac{2}{b}\frac{P}{h}s_{11} - \frac{2}{b\lambda_{33}}d_{31}d_{31}\frac{P}{h}]\frac{1}{h^2}x^3$$

$$+[-\frac{6}{b}\frac{M}{h^2}s_{11} - \frac{9}{20}s_{13}q - \frac{6}{10}s_{44}q - \frac{\lambda_{11}}{10(\lambda_{33})^2}d_{31}d_{31}q + \frac{9}{20\lambda_{33}}d_{31}d_{33}q - \frac{5}{5\lambda_{33}}d_{31}d_{15}q$$

$$+\frac{6}{b\lambda_{33}}d_{31}d_{31}\frac{M}{h^2} + \frac{1}{2\lambda_{11}}d_{15}d_{15}q]\frac{1}{h}x^2 - \{[2qs_{11} - \frac{2}{\lambda_{33}}d_{31}d_{31}q]\frac{l^3}{h^3} + [\frac{6}{b}\frac{P}{h}s_{11}$$

$$-\frac{6}{b\lambda_{33}}d_{31}d_{31}\frac{P}{h}]\frac{l^2}{h^2} + [-\frac{12}{b}\frac{M}{h^2}s_{11} - \frac{9}{10}s_{13}q - \frac{6}{5}s_{44}q - \frac{\lambda_{11}}{5(\lambda_{33})^2}d_{31}d_{31}q + \frac{9}{10\lambda_{33}}d_{31}d_{33}q$$

$$-\frac{2}{5\lambda_{33}}d_{31}d_{15}q + \frac{12}{b\lambda_{33}}d_{31}d_{31}\frac{M}{h^2} + \frac{1}{\lambda_{11}}d_{15}d_{15}q]\frac{l}{h}\}x + [\frac{3}{2}qs_{11} - \frac{3}{2\lambda_{33}}d_{31}d_{31}q]\frac{l^4}{h^3}$$

$$+[\frac{4}{b}\frac{P}{h}s_{11} - \frac{4}{b\lambda_{33}}d_{31}d_{31}\frac{P}{h}]\frac{l^3}{h^2} - [\frac{6}{b}\frac{M}{h^2}s_{11} + \frac{9}{20}s_{13}q + \frac{6}{10}s_{44}q + \frac{\lambda_{11}}{10(\lambda_{33})^2}d_{31}d_{31}q$$

$$-\frac{9}{20\lambda_{33}}d_{31}d_{33}q + \frac{1}{5\lambda_{33}}d_{31}d_{15}q - \frac{6}{b\lambda_{33}}d_{31}d_{31}\frac{M}{h^2} - \frac{1}{2\lambda_{11}}d_{15}d_{15}q]\frac{l^2}{h} \tag{A44}$$

and

$$
\begin{aligned}
u = {} & -2qs_{11}\tfrac{1}{h^3}x^3z - \tfrac{6}{b}\tfrac{P}{h^3}s_{11}x^2z + 2(2s_{13}+s_{44})q\tfrac{1}{h^3}xz^3 + \tfrac{12}{b}\tfrac{M}{h^3}s_{11}xz + \tfrac{3}{2}s_{13}q\tfrac{1}{h}xz \\
& -\tfrac{3(2s_{13}+s_{44})}{10}q\tfrac{1}{h}xz + \tfrac{2\lambda_{11}}{(\lambda_{33})^2}(d_{31})^2q\tfrac{1}{h^3}xz^3 + \tfrac{2(2s_{13}+s_{44})}{\lambda_{33}s_{11}}(d_{31})^2q\tfrac{1}{h^3}xz^3 - s_{13}\tfrac{q}{2}x \\
& -\tfrac{3\lambda_{11}}{10(\lambda_{33})^2}(d_{31})^2q\tfrac{1}{h}xz - \tfrac{3(2s_{13}+s_{44})}{10\lambda_{33}s_{11}}(d_{31})^2q\tfrac{1}{h}xz - \tfrac{4}{\lambda_{33}}d_{31}d_{33}q\tfrac{1}{h^3}xz^3 - \tfrac{1}{2\lambda_{33}}d_{31}d_{33}ql \\
& +\tfrac{3}{5\lambda_{33}}d_{31}d_{33}q\tfrac{1}{h}xz + \tfrac{4}{\lambda_{33}}d_{31}d_{15}q\tfrac{1}{h^3}xz^3 - \tfrac{3}{5\lambda_{33}}d_{31}d_{15}q\tfrac{1}{h}xz - 2s_{13}q\tfrac{1}{h^3}xz^3 \\
& +\tfrac{2}{\lambda_{33}}d_{31}d_{31}q\tfrac{1}{h^3}x^3z + \tfrac{6}{b\lambda_{33}}d_{31}d_{31}\tfrac{P}{h^3}x^2z - \tfrac{2\lambda_{11}}{(\lambda_{33})^2}d_{31}d_{31}q\tfrac{1}{h^3}xz^3 + \tfrac{1}{2\lambda_{33}}d_{31}d_{33}qx \\
& +\tfrac{\lambda_{11}}{2(\lambda_{33})^2}d_{31}d_{31}q\tfrac{1}{h}xz - \tfrac{2(2s_{13}+s_{44})}{\lambda_{33}s_{11}}d_{31}d_{31}q\tfrac{1}{h^3}xz^3 - \tfrac{12}{b\lambda_{33}}d_{31}d_{31}\tfrac{M}{h^3}xz \\
& +\tfrac{3(2s_{13}+s_{44})}{10\lambda_{33}s_{11}}d_{31}d_{31}q\tfrac{1}{h}xz + \tfrac{2}{\lambda_{33}}d_{31}d_{33}q\tfrac{1}{h^3}xz^3 - \tfrac{3}{2\lambda_{33}}d_{31}d_{33}q\tfrac{1}{h}xz + s_{13}\tfrac{q}{2}l \\
& -\tfrac{1}{\lambda_{33}}d_{31}d_{15}q\tfrac{1}{h}xz - \tfrac{2}{\lambda_{33}}d_{31}d_{15}q\tfrac{1}{h^3}xz^3 + \tfrac{3}{2\lambda_{33}}d_{31}d_{15}q\tfrac{1}{h}xz + [\tfrac{2}{b}s_{13}\tfrac{P}{h} + \tfrac{2}{b}s_{44}\tfrac{P}{h} \\
& -\tfrac{2}{b\lambda_{33}}d_{33}d_{31}\tfrac{P}{h} + \tfrac{2}{b\lambda_{33}}d_{15}d_{31}\tfrac{P}{h}]\tfrac{1}{h^2}z^3 + \{-\tfrac{3}{2b}s_{44}\tfrac{P}{h} - \tfrac{3}{2b\lambda_{33}}d_{15}d_{31}\tfrac{P}{h} + [2qs_{11} \\
& -\tfrac{2}{\lambda_{33}}d_{31}d_{31}q]\tfrac{l^3}{h^3} + [\tfrac{6}{b}\tfrac{P}{h}s_{11} - \tfrac{6}{b\lambda_{33}}d_{31}d_{31}\tfrac{P}{h}]\tfrac{l^2}{h^2} + [-\tfrac{12}{b}\tfrac{M}{h^2}s_{11} - \tfrac{9}{10}s_{13}q - \tfrac{6}{5}s_{44}q \\
& -\tfrac{\lambda_{11}}{5(\lambda_{33})^2}d_{31}d_{31}q + \tfrac{9}{10\lambda_{33}}d_{31}d_{33}q - \tfrac{2}{5\lambda_{33}}d_{31}d_{15}q + \tfrac{12}{b\lambda_{33}}d_{31}d_{31}\tfrac{M}{h^2} + \tfrac{1}{\lambda_{11}}d_{15}d_{15}q]\tfrac{l}{h}\}z
\end{aligned}
\tag{A45}
$$

Similarly, the expression of stress components can be obtained

$$
\begin{aligned}
\sigma_x = {} & -6q\tfrac{1}{h^3}x^2z - \tfrac{12}{b}\tfrac{P}{h^3}xz + [\tfrac{2(2s_{13}+s_{44})}{s_{11}}q\tfrac{1}{h^3} + \tfrac{2\lambda_{11}}{s_{11}(\lambda_{33})^2}(d_{31})^2q\tfrac{1}{h^3} - \tfrac{4}{\lambda_{33}s_{11}}d_{31}d_{33}q\tfrac{1}{h^3} \\
& +\tfrac{2(2s_{13}+s_{44})}{\lambda_{33}(s_{11})^2}(d_{31})^2q\tfrac{1}{h^3} + \tfrac{4}{\lambda_{33}s_{11}}d_{31}d_{15}q\tfrac{1}{h^3}]z^3 + [\tfrac{12}{b}\tfrac{M}{h^3} - \tfrac{3(2s_{13}+s_{44})}{10s_{11}}q\tfrac{1}{h} \\
& -\tfrac{3\lambda_{11}}{10s_{11}(\lambda_{33})^2}(d_{31})^2q\tfrac{1}{h} - \tfrac{3(2s_{13}+s_{44})}{10\lambda_{33}(s_{11})^2}(d_{31})^2q\tfrac{1}{h} + \tfrac{3}{5\lambda_{33}s_{11}}d_{31}d_{33}q\tfrac{1}{h} - \tfrac{3}{5\lambda_{33}s_{11}}d_{31}d_{15}q\tfrac{1}{h}]z
\end{aligned}
\tag{A46}
$$

$$
\sigma_z = -\frac{2q}{h^3}z^3 + \frac{3q}{2h}z - \frac{q}{2},
\tag{A47}
$$

and

$$
\tau_{zx} = \frac{6q}{h^3}xz^2 - \frac{3q}{2h}x + \frac{6P}{bh^3}z^2 - \frac{3P}{2bh}.
\tag{A48}
$$

The expressions of electric displacement components are

$$
D_x = \left(d_{15} + \frac{\lambda_{11}d_{31}}{\lambda_{33}}\right)\left(\frac{6q}{h^3}xz^2 - \frac{3q}{2h}x + \frac{6P}{bh^3}z^2 - \frac{3P}{2bh}\right)
\tag{A49}
$$

and

$$
\begin{aligned}
D_z = {} & [\tfrac{2\lambda_{11}}{s_{11}(\lambda_{33})^2}(d_{31})^3q\tfrac{1}{h^3} + \tfrac{2(2s_{13}+s_{44})}{\lambda_{33}(s_{11})^2}(d_{31})^3q\tfrac{1}{h^3} - \tfrac{4}{\lambda_{33}s_{11}}d_{31}d_{31}d_{33}q\tfrac{1}{h^3} - \tfrac{2\lambda_{11}d_{31}q}{\lambda_{33}h^3} \\
& +\tfrac{4}{\lambda_{33}s_{11}}d_{31}d_{31}d_{15}q\tfrac{1}{h^3} - \tfrac{2d_{15}q}{h^3}]z^3 + d_{31}[-\tfrac{3\lambda_{11}}{10s_{11}(\lambda_{33})^2}(d_{31})^3q\tfrac{1}{h} - \tfrac{3(2s_{13}+s_{44})}{10\lambda_{33}(s_{11})^2}(d_{31})^3q\tfrac{1}{h} \\
& +\tfrac{3}{5\lambda_{33}s_{11}}d_{31}d_{31}d_{33}q\tfrac{1}{h} - \tfrac{3}{5\lambda_{33}s_{11}}d_{31}d_{31}d_{15}q\tfrac{1}{h} + \tfrac{\lambda_{11}d_{31}q}{2\lambda_{33}h} + \tfrac{d_{15}q}{2h}]z
\end{aligned}
\tag{A50}
$$

References

1. Jiang, Y.G.; Gong, L.L.; Hu, X.H.; Zhao, Y.; Chen, H.W.; Feng, L.; Zhang, D.Y. Aligned P(VDF-TrFE) nanofibers for enhanced piezoelectric directional strain sensors. *Polymers* **2018**, *10*, 364. [CrossRef]
2. Elnabawy, E.; Hassanain, A.; Shehata, N.; Popelka, A.; Nair, R.; Yousef, S.; Kandas, I. Piezoelastic PVDF/TPU nanofibrous composite membrane: Fabrication and characterization. *Polymers* **2019**, *11*, 1634. [CrossRef]
3. Oh, W.J.; Lim, H.S.; Won, J.S.; Lee, S.G. Preparation of PVDF/PAR composites with piezoelectric properties by post-treatment. *Polymers* **2018**, *10*, 1333. [CrossRef]
4. Kim, M.; Wu, Y.S.; Kan, E.C.; Fan, J. Breathable and flexible piezoelectric ZnO@PVDF fibrous nanogenerator for wearable applications. *Polymers* **2018**, *10*, 745. [CrossRef]

5. Moghadam, A.; Kouzani, A.; Zamani, R.; Magniez, K.; Kaynak, A. Nonlinear large deformation dynamic analysis of electroactive polymer actuators. *Smart Struct. Syst.* **2015**, *15*, 1601–1623. [CrossRef]

6. Nasri-Nasrabadi, B.; Kaynak, A.; Komeily-Nia, Z.; Li, J.; Zolfagharian, A.; Adams, S.; Kouzani, A. An electroactive polymer composite with reinforced bending strength, based on tubular micro carbonized-cellulose. *Chem. Eng. J.* **2018**, *334*, 1775–1780. [CrossRef]

7. Gibeau, B.; Koch, C.R.; Ghaemi, S. Active control of vortex shedding from a blunt trailing edge using oscillating piezoelectric flaps. *Phys. Rev. Fluids* **2019**, *4*, 1–26. [CrossRef]

8. Moretti, M.; Silva, E.C.N.; Reddy, J.N. Topology optimization of flex tensional piezoelectric actuators with active control law. *Smart Mater. Struct.* **2019**, *28*, 1–16. [CrossRef]

9. Ji, H.L.; Qiu, J.H.; Wu, Y.P.; Zhang, C. Semi-active vibration control based on synchronously switched piezoelectric actuators. *Int. J. Appl. Electrom.* **2019**, *59*, 299–307. [CrossRef]

10. Wang, Z.K.; Chen, G.C. A general solution and the application of space axisymmetric problem in piezoelectric material. *Appl. Math. Mech.* **1994**, *15*, 587–598.

11. Lin, Q.R.; Liu, Z.X.; Jin, Z.L. A close form solution to simply supported piezoelectric beams under uniform exterior pressure. *Appl. Math. Mech.* **2000**, *21*, 617–624.

12. Mei, F.L.; Zeng, D.S. State equation method of mechanical-electric coupling for a piezoelectric beam. *J. Shandong Univ. Sci. Technol.* **2002**, *21*, 9–12.

13. Zhu, C.Z. Analytic solution to piezoelectric cantilever beam with concentrated force at free end. *J. Nanjing Inst. Technol.* **2001**, *1*, 12–15.

14. Ding, H.J.; Jiang, A.M. Polynomial solutions to piezoelectric beams(I)-several exact solutions. *Appl. Math. Mech.* **2005**, *26*, 1009–1015.

15. Ding, H.J.; Jiang, A.M. Polynomial solutions to piezoelectric beams(II)-Analytical solutions to typical problems. *Appl. Math. Mech.* **2005**, *26*, 1016–1021.

16. Ding, H.J.; Wang, G.Q.; Chen, W.Q. Green's functions for a two-phase infinite piezoelectric plane. *Proc. R. Soc.* **1997**, *453*, 2241–2257. [CrossRef]

17. Yang, D.Q.; Liu, Z.X. Analytical solution for bending of a piezoelectric cantilever beam under an end load. *Chin. Q. Mech.* **2003**, *24*, 327–333.

18. Pang, X.M.; Qiu, J.H.; Zhu, K.J.; Du, J.Z. (K, Na)NbO₃-based lead-free piezoelectric ceramics manufactured by two-step sintering. *Ceram. Int.* **2012**, *38*, 2521–2527. [CrossRef]

19. Zhu, Q.; Yue, J.Z.; Liu, W.Q.; Wang, X.D.; Chen, J.; Hu, G.D. Active vibration control for piezoelectricity cantilever beam: an adaptive feed forward control method. *Smart Mater. Struct.* **2017**, *26*, 047003. [CrossRef]

20. Peng, J.; Zhang, G.; Xiang, M.J.; Sun, H.X.; Wang, X.Y.; Xie, X.Z. Vibration control for the nonlinear resonant response of a piezoelectric elastic beam via time-delayed feedback. *Smart Mater. Struct.* **2019**, *28*, 1–12. [CrossRef]

21. Liu, Y.J.; Yang, D.Q. Analytical solution of the bending problem of piezoelectricity cantilever beam under uniformly distributed loading. *Acta Mech. Solida Sin.* **2002**, *23*, 366–372.

22. Huang, B.B.; Shi, Z.F. Several analytical solutions for a functionally gradient piezoelectric cantilever. *Acta Mater. Compos. Sin.* **2002**, *19*, 106–113.

23. Zhang, L.N.; Shi, Z.F. Analytical solution of simply-supported gradient piezoelectric beam. *J. North. Jiaotong Univ.* **2002**, *26*, 71–76.

24. Wang, Q.; Quek, S.T.; Sun, C.T.; Liu, X. Analysis of piezoelectric coupled circular plate. *Smart Mater. Struct.* **2001**, *10*, 229–239. [CrossRef]

25. Lian, Y.S.; He, X.T.; Shi, S.J.; Li, X.; Yang, Z.X.; Sun, J.Y. A multi-parameter perturbation solution for functionally graded piezoelectric cantilever beams under combined loads. *Materials* **2018**, *11*, 1222. [CrossRef] [PubMed]

26. Chen, S.L.; Li, Q.Z. The FPPM solutions for the problems of large deflection of axisymmetric circular plate. *J. Chongqing Jianzhu Univ.* **2003**, *25*, 32–36.

27. Lian, Y.S.; He, X.T.; Liu, G.H.; Sun, J.Y.; Zheng, Z.L. Application of perturbation idea to well-known Hencky problem: A perturbation solution without small-rotation-angle assumption. *Mech. Res. Commun.* **2017**, *83*, 32–46. [CrossRef]

28. Nowinski, J.L.; Ismail, I.A. Application of a multi-parameter perturbation method to elastostatics. *J Theor. App. Mech.* **1965**, *2*, 35–45.

29. Chien, W.Z. Second order approximation solution of nonlinear large deflection problem of Yongjiang Railway Bridge in Ningbo. *Appl. Math. Mech.* **2002**, *23*, 493–506.

Polymers **2019**, *11*, 1934

30. He, X.T.; Chen, S.L. Biparametric perturbation solutions of the large deflection problem of cantilever beams. *Appl. Math. Mech.* **2006**, *27*, 404–410. [CrossRef]
31. He, X.T.; Cao, L.; Li, Z.Y.; Hu, X.J.; Sun, J.Y. Nonlinear large deflection problems of beams with gradient: A biparametric perturbation method. *App. Math. Comput.* **2013**, *219*, 7493–7513. [CrossRef]
32. He, X.T.; Cao, L.; Sun, J.Y.; Zheng, Z.L. Application of a biparametric perturbation method to large-deflection circular plate problems with a bimodular effect under combined loads. *J. Math. Anal. Appl.* **2014**, *420*, 48–65. [CrossRef]
33. Ruan, X.P.; Danforth, S.C.; Safari, A.; Chou, T.W. Saint-Venant end effects in piezoceramic materials. *Int. J. Solids Struct.* **2000**, *37*, 2625–2637. [CrossRef]
34. He, X.T.; Yang, Z.X.; Jing, H.X.; Sun, J.Y. One-dimensional theoretical solution and two-dimensional numerical simulation for functionally-graded piezoelectric cantilever beams with different properties in tension and compression. *Polymers* **2019**, *11*, 1–24.

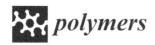

Article

Shape-Adaptive Metastructures with Variable Bandgap Regions by 4D Printing

Reza Noroozi [1,2], Mahdi Bodaghi [1,*], Hamid Jafari [2], Ali Zolfagharian [3] and Mohammad Fotouhi [4]

[1] Department of Engineering, School of Science and Technology, Nottingham Trent University, Nottingham NG11 8NS, UK; reza.noroozi@ut.ac.ir
[2] School of Mechanical Engineering, Faculty of Engineering, University of Tehran, Tehran 1417466191, Iran; hamid.jafari@ut.ac.ir
[3] School of Engineering, Deakin University, Geelong, VIC 3216, Australia; a.zolfagharian@deakin.edu.au
[4] School of Engineering, University of Glasgow, Glasgow G12 8QQ, UK; mohammad.fotouhi@glasgow.ac.uk
* Correspondence: mahdi.bodaghi@ntu.ac.uk; Tel.: +44-115-84-83470

Received: 13 January 2020; Accepted: 24 February 2020; Published: 1 March 2020

Abstract: This article shows how four-dimensional (4D) printing technology can engineer adaptive metastructures that exploit resonating self-bending elements to filter vibrational and acoustic noises and change filtering ranges. Fused deposition modeling (FDM) is implemented to fabricate temperature-responsive shape-memory polymer (SMP) elements with self-bending features. Experiments are conducted to reveal how the speed of the 4D printer head can affect functionally graded prestrain regime, shape recovery and self-bending characteristics of the active elements. A 3D constitutive model, along with an in-house finite element (FE) method, is developed to replicate the shape recovery and self-bending of SMP beams 4D-printed at different speeds. Furthermore, a simple approach of prestrain modeling is introduced into the commercial FE software package to simulate material tailoring and self-bending mechanism. The accuracy of the straightforward FE approach is validated against experimental observations and computational results from the in-house FE MATLAB-based code. Two periodic architected temperature-sensitive metastructures with adaptive dynamical characteristics are proposed to use bandgap engineering to forbid specific frequencies from propagating through the material. The developed computational tool is finally implemented to numerically examine how bandgap size and frequency range can be controlled and broadened. It is found out that the size and frequency range of the bandgaps are linked to changes in the geometry of self-bending elements printed at different speeds. This research is likely to advance the state-of-the-art 4D printing and unlock potentials in the design of functional metastructures for a broad range of applications in acoustic and structural engineering, including sound wave filters and waveguides.

Keywords: 4D printing; metastructure; shape-memory polymers; wave propagation; finite element method; bandgap

1. Introduction

In order to survive in variable environments and keep their performance, natural materials have evolved to be active and adaptive, retaining functions across a range of stresses or strains or changing thermomechanical properties in response to external stimuli like light, temperature or moisture [1]. Several active materials have been developed to mimic the unique properties of natural materials and adaptive structures [2]. Among these active materials, shape-memory polymers (SMPs) have attracted much attention due to their lower density, higher recoverable strain of up to 400%, lower cost, simple shape programming procedure, and excellent controllability over the recovery temperature [3,4].

In the recent two decades, three-dimensional (3D) printing technology, also known as additive manufacturing (AM), has gained considerable attention as an advanced manufacturing technique that can create complex objects through depositing materials in a layer-by-layer manner [5–9]. With the introduction of active materials, 3D printing approaches have shown excellent potential for the fabrication of adaptive structures, namely four-dimensional (4D) printed structures, with the capability of reshaping their configuration and changing their properties over time [10–12]. For the first time, Tibbits [13] experimentally demonstrated how 4D-printed objects could transform over time and perform self-assemblies. While 3D printing methods can be used to fabricate static structures, 4D printing methods allow the fabrication of dynamically reconfigurable architectures with desired functionality and responsiveness. Considering a specific application, 4D-printed objects can be designed to respond to environmental stimuli and external triggers such as humidity, light, heat, and electric or magnetic fields [14–16]. For example, by 4D printing temperature-sensitive SMPs, Bodaghi et al. [17] introduced adaptive metamaterials with the capability of 1D/2D-to-2D/3D shape-shifting through self-folding and/or self-coiling. Wang et al. [18] introduced a novel 4D printing technology of composites with continuous embedded fibers for which the programmable deformation was caused by the difference in coefficient of thermal expansion between continuous fibers and flexible matrix. Zhang et al. [19] 4D-printed lightweight structures with self-folding/unfolding performances when exposed to a certain temperature. Also, Zhao et al. [20] 4D-printed SMPs by stereolithography of photopolymers. Fold-deploy test and shape-memory cycle measurements proved a high shape recovery rate, shape fixity, and recovery of the printed objects. Zolfagharian et al. [21] achieved controlled bending in a commercial prestrained SMP film by using different 4D-printed patterns and a number of layers. The photothermal stimulus was used to induce differential shrinking through the thickness of the actuator hinge. Recently, Wu et al. [10] designed a new 4D printable acrylate-based photosensitive resin prepared for digital light processing. They also investigated the influence of crosslinker concentration on the shape memory and mechanical properties.

Acoustic metamaterials are a new class of architected materials designed to control, direct, and manipulate waves through the material. Wave propagation in periodic metamaterials has been investigated broadly. Research works have revealed that these architected structures can be designed to gain desirable dispersion performance [22], lightweight structures [23], and excellent energy absorbers [24] by arranging unit cells in periodic repetition. The dispersion performance may present stop-band and pass-band frequency ranges as well. Phani et al. [25] introduced a computational procedure to find bandgaps by using Floquet–Bloch principles. Matlack et al. [26] fabricated metastructures by combining 3D printing polycarbonate lattices with steel cubes assembling. They showed that the value of bandgap could be varied by changing the lattice geometry and structural stiffness. By using deployable structures composed of beams and torsion springs, Nadda and Karami [27] showed that wave transmission characteristics could be tuned by modulating the fold angle. By combining a periodic lattice and locally-resonant inclusions with different temperature dependency, Nimmagadda and Matlack [28] proposed a metastructure that can tune the bandgaps under thermal stimuli.

The main objective of this research is to show how 4D printing technology can be used to engineer adaptive metastructures with the ability to control elastic wave propagation. Such structure is essential in vibration mitigation and acoustic attenuation. Inspired by thermomechanics of SMPs and the potential of fused deposition modeling (FDM) in 4D printing self-bending elements, adaptive functionally graded (FG) beams are fabricated. It is shown, experimentally and numerically, how 4D printing speed can control shape recovery and self-bending features of active elements. A 3D constitutive model, along with an in-house finite element (FE) code programmed in MATLAB software, is developed to replicate shape recovery of FG active elements 4D-printed at four different printing speeds. Afterward, a simple approach, along with the commercial FE software package of COMSOL Multiphysics, is introduced to simulate the material tailoring in the 4D printing stage and self-bending of active SMP elements. The accuracy and simplicity of the straightforward FE approach implemented

Polymers **2020**, *12*, 519

in commercial FE software are assured via comparative studies with experimental data and numerical results from the in-house FE MATLAB-based code. Two periodic architected temperature-sensitive metastructures with adaptive dynamical characteristics are conceptually proposed. Their dynamic behaviors during heating–cooling processes are numerically studied in detail. Adaptive wave propagation (pass-band and stop-band), desired dynamic performance, and vibration manipulation are numerically demonstrated to be some of the unique characterizations of these metastructures. Computational studies reveal that, while the bandgaps in these metastructures are induced by the self-bending mechanism, their crucial feature is that the bandgap size and frequency range can be controlled and broadened through local resonances linked to changes in the structural geometry. The material/structural formulation, concepts, and results provided in this article are expected to be instrumental towards 4D printing adaptive metastructures for a broad range of applications, including structural vibration absorption, waveguiding, and noise mitigation.

2. Conceptual Design

2.1. Four-Dimensional Printing SMPs

In this section, by understanding the shape-memory effect (SME) and FDM technology, the FG 4D printing concept for designing adaptive structures is introduced. Temperature-sensitive SMPs are a class of smart materials that can recover their original shape from a temporary programmed shape by heating. In the programming step, the material, initially in a strain/stress-free state at a temperature lower than the transition temperature ($T < T_g$), is firstly heated up to T_h that is higher than transition temperature ($T_g < T_h$). The material that is stable at the rubbery phase is then loaded and held fixed while being cooled down to T_l, which is less than the transition temperature ($T_l < T_g$). By removing the constraints, an inelastic strain, so-called prestrain, remains in the material and forms an irregular shape. The material is in a free-stress state at this stage. In the shape recovery process, the SMP is heated to recover its original shape, which is known as free strain recovery, and finally is cooled back to the low temperature.

FDM technology, as a filament-based material-extrusion 3D printing method, applies a similar thermomechanical process on the material during the fabrication. Therefore, it may have the potential to fabricate 4D SMP architectures along with the shape programming. Figure 1 depicts a schematic of FDM technology. At first, the material is heated inside the liquefier up to T_{ln}, which is higher than the transition temperature (T_g), and then forced out of the nozzle and deposited onto the platform by the 4D printer head moving at speed S_p. In this step, the material is stretched similar to the heating–loading process of the SMP programming step that induces the prestrain. Therefore, the printing speed may affect the prestrain value. It would be sensible that greater speed produces more significant mechanical loading, hence inducing greater prestrain. After deposition, the printed layer cools and solidifies in the same manner as the cooling step in the programming process. After 4D printing the layer, the platform moves downward, and the 4D printer head proceeds to deposit the following layer. The programming procedure is finalized by mechanical unloading through removing the printed object from the build tray.

The thermal/surface boundary condition between the 4D-printed layers may affect the through-the-thickness prestrain regime. For instance, while the first printing layer is deposited on the stiff and rough build tray, other layers above it are laid on the previously printed polymeric layers. Therefore, material and surface conditions may affect bonding and stretching conditions, reducing the first layer prestrain. The first layer is expected to show the lowest prestrain value. By printing the second layer, the first layer is partially reheated, and this extra heat may reduce the prestrain value. In other words, the first layer and layers above it, except the end layer, are always reheated, and their prestrain value is decreased. Since the last layer never gets any extra heat, it is expected to have the maximum prestrain. It may be concluded that the prestrain regime may have an increasing trend through the thickness upward from the lower to the upper layer. This additive manufacturing process can be called FG 4D printing as the material is programmed during the fabrication in the same manner as an FG material.

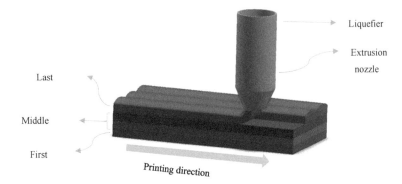

Figure 1. A schematic of the fused deposition modeling (FDM) method.

2.2. Material Behaviors

In this research, we used the polylactic acid (PLA) filament with a diameter of 1.75 mm and glass transition temperature of 65 °C. Objects were fabricated using a 3Dgence DOUBLE printer developed by 3Dgence. This is a low-cost desktop 3D printer that extrudes 1.75 mm filaments with a 0.4 mm nozzle. The 3Dgence Slicer software was used to produce G-code files from STL files and command and control parameters of liquefier temperature and printing speed. Beam-like elements were 4D-printed with dimensions of (30 × 1.6 × 1) mm for length, width, and thickness, respectively. Each printing layer was considered to have a 0.1 mm thickness. The print raster was assumed to be along the length direction. The temperature of the liquefier, build tray, and chamber were set 210, 24, and 24 °C, respectively. In all the 4D printings, unless otherwise stated, the printing speed was set to 5 mm/s in which the printer does not induce any prestrains in the materials.

Dynamic-mechanical analyzer (DMA, Q800, TA Instruments, New Castle, DE, USA) was first implemented to determine the temperature-dependent mechanical properties of 3D-printed PLAs. DMA test was conducted in an axial tensile way with the frequency of force oscillation 1 Hz and heating rate 5 °C/min ranging from 30 to 93 °C. The applied dynamic stress was 1.5 times the static stress. Further, the thermomechanical behavior of the 4D-printed sample in terms of storage modulus, E_s, and phase lag, $\tan(\delta)$, is shown in Figure 2. It is seen that the storage modulus is reduced drastically as the material is heated to 60 °C. It is also observed that the phase lag graph peaks at 65 °C, which is assumed to be the glass transition temperature.

Five beam-like elements were 4D-printed at five different speeds (Sp = 5, 10, 20, 40, 70 mm/s). Figure 3 shows the beam configuration after 4D printing. After 4D printing, five samples were heated by dipping into the hot water with a prescribed temperature of 85 °C (20 °C higher than the transition temperature) and then were cooled down to the room temperature, 24 °C. Figure 4 depicts the sample configuration after the heating–cooling process. As can be seen, samples with a straight temporary shape may transform into curved beams. This means that the samples may already be programmed and prestained during the 4D printing process. Figure 4a shows that the sample printed at a low speed of 5 mm/s does not undergo any changes by heating. This implies that this printing speed is not enough to produce any prestrains. However, the configuration presented in Figure 4b indicates that increasing the printing speed to 10 mm/s can cause a slight curvature after heating. Thus, this speed can be considered as a transient speed. The configuration change could be due to an unbalanced FG prestrain regime induced through the thickness direction during the 4D printing process. When dipping the samples into hot water, the unbalanced prestrain, with an increasing trend through the thickness upward from the lower layer to the upper layer, was recovered, leading to a self-bending movement. The mismatch in free-strain recovery inducing curvatures enabled the overall configuration to change toward the top layer. It is worthwhile to mention that an increase in the 4D printing speed increased the bending angle and curvature. The faster the 4D printing, the greater the prestrain and consequently the deformation. Finally, experiments revealed that the FDM 4D printing technology has high potential in fabricating and programming adaptive objects with self-bending features.

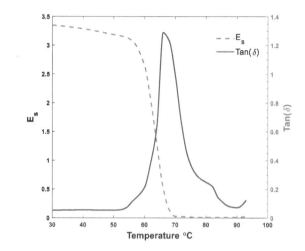

Figure 2. Dynamic-mechanical analyzer (DMA) test for the 3D-printed polylactic acid (PLA).

Figure 3. The beam configuration after 4D printing.

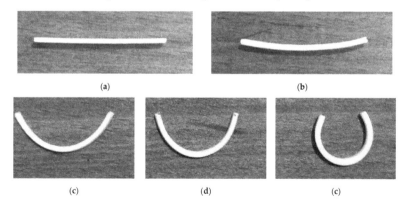

Figure 4. The configuration of the samples 4D-printed at different speeds of (**a**) 5, (**b**) 10, (**c**) 20, (**d**) 40, and (**e**) 70 mm/s after the heating–cooling process.

3. Theoretical Modeling

The 4D-printed metastructures with adaptive wave propagation feature can be designed based on the theoretical understanding of the FG programming process and SME of 4D-printed SMP elements. This section is devoted to developing constitutive models and mathematical formulation to predict SME and wave propagation in 4D-printed SMPs.

3.1. SMP Model

In this division, a phenomenological constitutive model presented initially in [17,29] is reformulated to describe shape programming and recovery processes in the 4D-printed structures. Analytical straightforward closed-form solutions are also derived.

The SMPs consist of two phases, so-called glassy and rubbery phases, which are stable at temperatures above and below the transition temperature, respectively. As SMPs show a mixture of glassy and rubbery phases, volume fractions of the rubbery and glassy phases describe the state of SMPs. The following scalar variables characterize them:

$$\eta_g = \frac{V_g}{V} \qquad \eta_r = \frac{V_r}{V} \tag{1}$$

where V_g and V_r indicate the volume of the glassy and rubbery phases, respectively. The subscripts 'g' and 'r' stand for glassy and rubbery phases, respectively, here and henceforth. The change between these phases is considered to be only a function of temperature as a generally well-known assumption. It implies that η_g and η_r are just functions of the temperature. Because in every moment the summation of these parameters must satisfy the unity ($\eta_g + \eta_r = 1$), the volume fraction of the rubbery phase can be expressed in terms of the glassy one as:

$$\eta_r(T) = 1 - \eta_g(T) \tag{2}$$

Considering experimental results from the DMA test, η_g can explicitly be interpolated by a hyperbolic function as [17,29]:

$$\eta_g = \frac{\tanh(a_1 T_g - a_2 T) - \tanh(a_1 T_g - a_2 T_h)}{\tanh(a_1 T_g - a_2 T_h) - \tanh(a_1 T_g - a_2 T_l)} \tag{3}$$

in which a_1 and a_2 are determined by adjusting the DMA curve.

The glassy and rubbery phases in SMPs are assumed to be linked in series, i.e., $\sigma = \sigma_g = \sigma_r$, where σ denotes the stress. Considering the fact that the 4D-printed objects may experience small strains and moderately large rotations, a small strain regime is assumed. Consequently, an additive strain decomposition is adopted as:

$$\varepsilon = \eta_g \varepsilon_g + (1 - \eta_g)\varepsilon_r + \varepsilon_{in} + \varepsilon_{th} \tag{4}$$

where ε denotes the total strain while ε_g and ε_r designate the strains of the glassy and rubbery phases, respectively. Also, ε_{in} is the inelastic strain due to the SMP phase transformation, while ε_{th} indicates the thermal strain that can be expressed as $\varepsilon_{th} = \int_{T_0}^{T}\left(\alpha_r + (\alpha_g - \alpha_r)\eta_g(T)\right)dT$ in which α_r and α_g denote thermal expansion in rubbery and glassy phases, respectively, and T_0 is the reference temperature.

Next, ε_{in}, associated with the glassy–rubbery phase transformation mechanism, will be formulated. During the cooling step, the rubbery phase changes into the glassy one, and its strain is stored in the SMP material. The strain storage is assumed to be proportional to η_g based on the rubbery phase strain. In the heating step, the stored strain is assumed to be released gradually, proportional to η_g concerning the preceding glassy phase and strain storage. Therefore, ε_{in} can mathematically be formulated in a rating format as:

$$\dot{\varepsilon}_{in} = \begin{cases} \dot{\eta}_g \, \varepsilon_r & \dot{T} < 0 \\ \frac{\eta_g}{\eta_g} \, \varepsilon_{in} & \dot{T} > 0 \end{cases} \tag{5}$$

By considering ε and T as external control variables and ε_g, ε_r, ε_{in}, and η_g as internal variables, introducing Helmholtz free energy density functions, implementing the second law of thermodynamics in the sense of the Clausius–Duhem inequality, and following a standard argument [30], the stress state can be obtained as:

$$\sigma = \sigma_g = \sigma_r \tag{6}$$

This equation is consistent with the taken assumption that the stress in each phase is equal in the series model. The glassy- and rubbery-phase stresses are also derived as:

$$\sigma_g = C_g \varepsilon_g, \, \sigma_r = C_r \varepsilon_r \tag{7}$$

in which C signifies the elasticity stiffness matrix defined as:

$$C = \frac{E}{(1+v)(1-2v)} \begin{bmatrix} 1-v & v & v & 0 & 0 & 0 \\ v & 1-v & v & 0 & 0 & 0 \\ v & v & 1-v & 0 & 0 & 0 \\ 0 & 0 & 0 & \frac{(1-2v)}{2} & 0 & 0 \\ 0 & 0 & 0 & 0 & \frac{(1-2v)}{2} & 0 \\ 0 & 0 & 0 & 0 & 0 & \frac{(1-2v)}{2} \end{bmatrix} \tag{8}$$

where E and v denote Young's modulus and Poisson's ratio, respectively. The stress–strain relationship can be derived by substituting Equation (7) into Equation (4) as:

$$\sigma = (S_r + \eta_g(S_g - S_r))^{-1}(\varepsilon - \varepsilon_{in} - \varepsilon_{th}) \tag{9}$$

where S indicates the compliance matrix defined as C^{-1}.

From a numerical viewpoint, the nonlinear SMP behaviors can be treated in an explicit time-discrete stress/strain-temperature-driven framework. The time domain $[0, t]$ is discretized into increments, and the evolution equation is solved over the local band $[t^n, t^{n+1}]$. The superscript $n + 1$ shows the current step, while the superscript n denotes the previous step. The inelastic strain can be computed by applying the linearized implicit backward-Euler integration method to the flow rule (5) as:

$$\varepsilon_{in}^{n+1} = \begin{cases} \varepsilon_{in}^n + \Delta\eta_g^{n+1} \, \varepsilon_r^{n+1} & \dot{T} < 0 \\ \varepsilon_{in}^n + \frac{\Delta\eta_g^{n+1}}{\eta_g^{n+1}} \, \varepsilon_{in}^{n+1} & \dot{T} > 0 \end{cases} \tag{10}$$

where

$$\Delta\eta_g^{n+1} = \eta_g^{n+1} - \eta_g^n \tag{11}$$

By using Equations (7) and (9) to substitute the rubbery strain and the stress, and performing some mathematical simplifications, ε_{in} defined in (10) can be updated for cooling and heating steps in stress and strain control ways as:

Cooling:

$$\text{stress control mode} \rightarrow \varepsilon_{in}^{n+1} = \varepsilon_{in}^n + \Delta\eta_g^{n+1} S_r^{n+1} \sigma^{n+1} \tag{12}$$

$$\text{strain control mode} \rightarrow \varepsilon_{in}^{n+1} = (I + \Delta\eta_g^{n+1} S_r^{n+1} C_e^{n+1})^{-1} (\varepsilon_{in}^n + \Delta\eta_g^{n+1} S_r^{n+1} C_e^{n+1} (\varepsilon^{n+1} - \varepsilon_{th}^{n+1})) \tag{13}$$

Heating:

$$\varepsilon_{in}^{n+1} = \frac{\eta_g^{n+1}}{\eta_g^n} \varepsilon_{in}^n \tag{14}$$

Finally, by considering updated inelastic strain, the stress–strain constitutive Equation (9) can be discretized and unified for heating and cooling processes as:

$$\sigma^{n+1} = C_D^{n+1}\left(\varepsilon^{n+1} - \varsigma\varepsilon_{in}^{n+1} - \varepsilon_{th}^{n+1}\right) \tag{15}$$

where the so-called unified stiffness matrix C_D and parameter ς for cooling and heating processes are defined as:

$$\begin{cases} C_D^{n+1} = \left(I + \Delta\eta_g^{n+1} S_r^{n+1} C_e^{n+1}\right)^{-1} C_e^{n+1}, \varsigma = 1 & \dot{T} < 0 \\ C_D^{n+1} = C_e, \varsigma = \frac{\eta_g^{n+1}}{\eta_g^n} & \dot{T} > 0 \end{cases} \tag{16}$$

3.2. Wave Propagation Model

The wave propagation analysis of the architected periodic structures can be carried out by using the Bloch's theorem for local resonance. Based on this theory, the displacement of each node in a chosen unit cell in the region of a periodic structure depends only on the displacement field of the equal node in the reference unit cell ($\vec{U}_{Ref}(\vec{r})$). The formulation of this theorem is implemented to solve the equations of motion in the periodic boundary conditions. This concept is stated as [31]:

$$\vec{U}\left(\vec{r} + \vec{R}, t\right) = \vec{U}_{Ref}\left(\vec{r}\right)\exp\left(i\left[\vec{\kappa}.\left(\vec{r} + \vec{R}\right) - \omega t\right]\right) \tag{17}$$

where $\vec{r}$ and $\vec{R}$ are position and lattice vectors, respectively. Also, the Bloch wave vector $\vec{\kappa}$ in the 2D periodicity is considered as $\vec{\kappa} = \left(\kappa_x, \kappa_y\right)$, where κ_x and κ_y denote the phase constants which are measures of the phase variations over one unit cell in two directions of the periodicity (X–Y). Also, ω and t refer to frequency and time, respectively. As illustrated in Figure 5, the periodicity in two directions is defined by the direct vectors $\vec{a}_x$ and $\vec{a}_y$. Hence, the lattice vector is described as:

$$\vec{R} = n_x \vec{a}_x + n_y \vec{a}_y \quad , \quad n_x, n_y = 0, \pm 1, \pm 2, \cdots \tag{18}$$

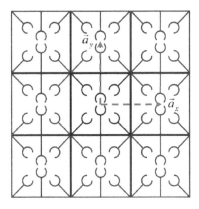

Figure 5. Direct vectors in a periodic model.

Since the Bloch wave vector varies in the Brillouin zone, the procedures to calculate reciprocal vectors and the first Brillouin zone are explained. Equation (19) formulates the reciprocal vectors $\vec{b}_x$ and $\vec{b}_y$ as [32]:

$$\begin{cases} \vec{b}_x = \dfrac{2\pi}{\left|\vec{a}_x \times \vec{a}_y\right|} \begin{bmatrix} a_{yY} \\ -a_{yX} \end{bmatrix} \\ \vec{b}_y = \dfrac{2\pi}{\left|\vec{a}_x \times \vec{a}_y\right|} \begin{bmatrix} -a_{xY} \\ a_{yX} \end{bmatrix} \end{cases} \tag{19}$$

where (a_{xX}, a_{xY}) and (a_{yX}, a_{yY}) are components of direct vectors $\vec{a}_x$ and $\vec{a}_y$ along the X- and Y-directions, respectively. The first Brillouin zone (FBZ) and irreducible Brillouin zone (IBZ) are depicted in Figure 6. The IBZ is the smallest part of the FBZ representing all the high-symmetry points in which the wave propagation is analyzed [33].

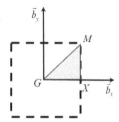

Figure 6. First Brillouin zone (FBZ, dashed square) and irreducible Brillouin zone (IBZ, colored triangle).

In this research, numerical simulations have been produced by employing an FE software. Each unit cell is discretized into finite elements. The dispersion relation is an eigenvalue problem that is solved By COMSOL Multiphysics 4.3. The problem has three unknown parameters (κ_x, κ_y, and ω) in which the wave vectors are examined in the IBZ (G-X-M-G), and the problem is determined to calculate the eigenfrequencies. Convergence studies of the mesh size have also been performed to achieve the converged results accurately to three significant digits.

3.3. FE Solution

3.3.1. In-House FE Code

In order to replicate thermomechanical behaviors of FG 4D-printed elements, a Ritz-based FE solution is developed in MATLAB software. A 3D 20-node quadratic serendipity hexahedron element is considered in this problem. It has 20 nodes so that 8 corner nodes are augmented with 12 side nodes located at the side center. The element also has 3 degrees of freedom per node (u_i, i = 1, 2, 3). More details on the FE solution and numerical programming can be found in [17].

3.3.2. COMSOL Multiphysics FE Modeling

In this section, thermomechanical behaviors of 4D-printed samples are analyzed by implementing a simple method in COMSOL Multiphysics software. For this purpose, by using the DMA test data, the temperature-dependent Young's modulus is implemented in the COMSOL Multiphysics. Table 1 shows the dependency of Young's modulus on the temperature.

Table 1. The temperature-dependent Young's modulus from the DMA test.

$T\,(^{\circ}C)$	30	40	50	60	70	80	90
$E\,(MPa)$	3350	3280	3166	2554	48	18	14

As described already, during 4D printing, a prestrain that varies through the thickness is induced in the object producing an FG structure. For modeling 4D-printed structures in COMSOL Multiphysics, the object can be divided into multiple sections with variable thermal expansion. In this study, the 4D-printed beam-like structures are divided into six sections. Figure 7 illustrates a discretized form of a printed beam with different thermal expansion coefficients.

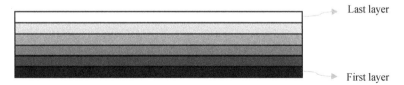

Last layer

First layer

Figure 7. Discretized 4D-printed sample.

The thermal expansion of each layer is chosen to replicate a configuration similar to the experiments as depicted in Figure 4 for a specific printing speed. Table 2 indicates the thermal expansion coefficient assumed for each layer.

Table 2. The thermal expansion coefficient of each layer for different printing speeds.

α_i (1/°C)	S_p (mm/s)			
	10	20	40	70
α_1	−0.0006	−0.0016	−0.0018	−0.00252
α_2	−0.0004	−0.0014	−0.0016	−0.00222
α_3	−0.0002	−0.0011	−0.0013	−0.0022
α_4	−0.00009	−0.0008	−0.0011	−0.00172
α_5	−0.00007	−0.0006	−0.0008	−0.00152
α_6	−0.00005	−0.0004	−0.0005	−0.00122

A relationship between thermal expansion coefficients and printing speed can be formulated as:

$$\alpha_i = C_1 S^3{}_P + C_2 S^2{}_P + C_3 S_P + C_4 \qquad 10 \leq S_P \leq 70 \qquad (20)$$

where C1, C2, and C3 are constants defined for each layer in Table 3.

Table 3. The constant of thermal expansion interpolation function for each printing layer.

Layer	Coefficient				
	$C_1 (10^{-8})$	$C_2 (10^{-6})$	$C_3 (10^{-3})$	$C_4 (10^{-3})$	$R - Square$
1	−3.1018	4.125	−0.1845	081	0.9722
2	−3.6	4.739	−0.2039	1.16	0.9824
3	−3.44	4.331	−0.1834	1.198	0.9864
4	−2.2	2.917	−0.135	0.9653	0.9917
5	−1.744	2.159	−0.0097	0.6798	0.989
6	−0.804	0.8966	−0.0051	0.2867	0.9716

Figure 8 shows the deformed configuration obtained from the FE COMSOL Multiphysics simulation of self-bending 4D-printed beams after the heating–cooling process. In order to characterize the configuration of the printed samples after the heating–cooling process, three geometric parameters are considered. For this purpose, we use parameters $R_1, R_2,$ and R_3 which describe the outer length, opening, and depth of mid surface, respectively, as shown in Figure 8c. In order to determine the accuracy and efficiency of the simple method implemented in COMSOL Multiphysics, the geometric features obtained from the experiments, FE COMSOL Multiphysics, and in-house FE code are compared in Table 4. It can be concluded that the simulation results of FE COMSOL Multiphysics are in good agreement with those measured from experiments and calculated by the in-house FE solution. This way, the reliability and accuracy of setting variable thermal expansion coefficients in the COMSOL Multiphysics in replicating the self-bending feature observed in the 4D-printed samples is validated.

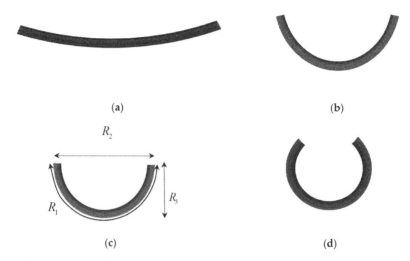

Figure 8. Finite element (FE) COMSOL Multiphysics simulation of the samples 4D-printed with different speeds of (**a**) 10, (**b**) 20, (**c**) 40, and (**d**) 70 mm/s after the heating–cooling process.

Table 4. The geometric parameters of the beam-like 4D structures after the heating–cooling process.

Method	S_p (mm/s)	$R_1 (mm)$	$R_2 (mm)$	$R_3 (mm)$
Experiment	10	29.8	28.2	3.1
	20	29.3	19	8.3
	40	29.1	16.3	9.2
	70	29.0	7.1	10.5
FE COMSOL Multiphysics	10	29.7	28.3	3.2
	20	29.4	19.1	8.4
	40	29.2	16.2	9.1
	70	28.9	7.0	10.4
In-house FE method	10	29.9	28.3	3.0
	20	29.1	19.2	8.4
	40	29.4	16.3	9.2
	70	29.2	7.2	10.3

3.4. Periodic Structural Design

In this section, two periodic architected metastructures with adaptive dynamical characteristics are conceptually proposed. These metastructures are made of passive mainframes printed at a low speed so that no prestrain is induce, and active beam-like members printed at high speed with induced prestrains and self-folding features. They have the potential to be 4D-printed by setting different printing speeds for two nozzles. The PLA material as characterized in Sections 2.2 and 3.3 are considered for 4D printing. The passive main frame of the metastructure is printed at a low speed such as 5 mm/s, while active elements with self-bending features are fabricated with three different 4D printing speeds. Different arrangements have been designed, and their dynamic performance has been examined numerically. Two metastructures have shown a high dynamic performance that will further be examined numerically in the next section. The first adaptive structure consists of active elements connected in parallel and diagonal to the frame, while the second adaptive structure consists of active elements connected in parallel with the frame. For convenience, they are called diagonal and parallel metastructures, respectively. Figure 9 shows the designed structures in which yellow and blue colors signify active and passive elements, respectively.

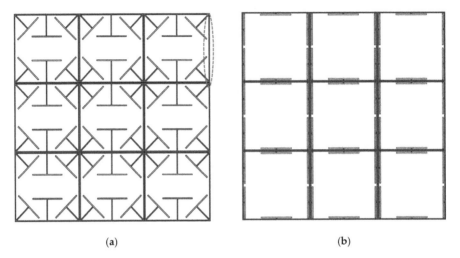

(a)   (b)

Figure 9. Periodic metastructures with active and passive components: (a) diagonal structure; (b) parallel structure (the red dashed oval shows the fixed-fixed beam used for the frequency normalization).

4. Results and Discussion

Periodic active structures are associated with granting propagating and bandgap ranges. In the propagating frequency ranges, the elastic wave propagation is done in all directions, but the bandgap represents a frequency range where the elastic wave propagation is stopped. In this section, COMSOL-based numerical results are presented, revealing how to design adaptive periodic structures with the ability to optimize the dynamical functionality without embedding any additional resonating components. Since it is difficult to actuate all mode shapes in experimental studies, experimental works have been considered in-plane or out-of-plane mode shapes only. However, in the present numerical study, a 3D dynamic case is investigated by FE COMSOL Multiphysics, and all modes of vibrations (i.e., bending, torsion, and elongation) can be measured without any limitation.

In order to verify the FE simulation, the bandgap of a triangular structure is simulated, as shown in Figure 10, and compared with the results of [25]. It is seen that there is an excellent agreement between the dispersion curves from the FE simulation and [25].

Eigenfrequencies of adaptive periodic structures, as shown in Figure 9, are computed by imposing periodic boundary conditions in different elastic wave vectors which are estimated based on IBZ detailed in the previous section. These eigenfrequencies are finally normalized against $\Omega = \omega / \omega_0$, in which $\omega_0 = 22.4 \sqrt{EI/(mL_0^4)}$ is the first natural frequency of a fixed-fixed single beam, as shown in Figure 9a. The dispersion diagram and some out-of-plane mode shapes of the diagonal metastructure are illustrated in Figure 11, whereas the mode shapes are depicted in high symmetry points (G, X, or M). As can be seen, there is a wide bandgap (gray square) in the range of $\Omega = 1.902 - 2.043$ which is 4.68% of the frequency ranging between 0 and 3. As shown in Figure 11, the cause of bandgap in this range is the resonance of active parts. Further, there is a flat eigenfrequency before the bandgap which helps to assure its local resonance nature.

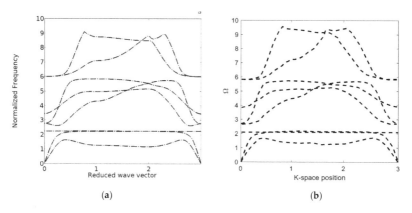

Figure 10. Verification of the band structure of the triangular topology: (**a**) the current study; (**b**) Ref. [25].

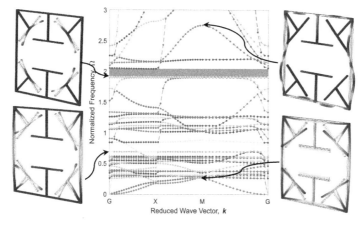

Figure 11. Band structure and mode shapes of the diagonal metastructure.

The configuration of the diagonal structure after the heating–cooling process for three printing speeds of active elements is illustrated in Figure 12. As it is expected, the elements 4D-printed faster produce more curvature after thermal activation. It is worth mentioning that the heating could also be another variable to control the curvature. i.e., by partially heating the metastructure with active elements printed at 70 mm/s speed, configurations become like those of thoroughly heated structures with active elements printed at 20 and 40 mm/s. Dispersion curves of diagonal structure with active elements of different 4D printing speeds after the heating–cooling process are depicted in Figures 13–15. Comparing the results presented in Figures 11 and 13 reveal that the bandgap area is adapted from $\Omega = 1.902 - 2.043$ to $\Omega = 1.751 - 1.812$, and the expanse of the bandgap area decreases from 4.68% to 2.01% as well. This phenomenon shows the significant effect of the self-bending feature on the locally resonant filtration. As can be seen, the functional range changes by tuning the 4D printing speed of active elements.

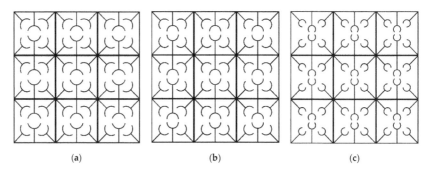

Figure 12. The configuration of adaptive periodic diagonal metastructure after heating–cooling process for three different printing speeds: (**a**) 20, (**b**) 40, and (**c**) 70 mm/s.

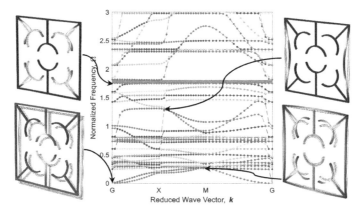

Figure 13. Band structure and mode shapes of the diagonal structure with active elements printed at 20 mm/s after the heating–cooling process.

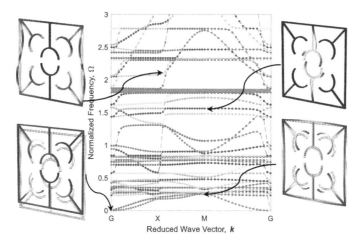

Figure 14. The counterpart of Figure 12 for 40 mm/s.

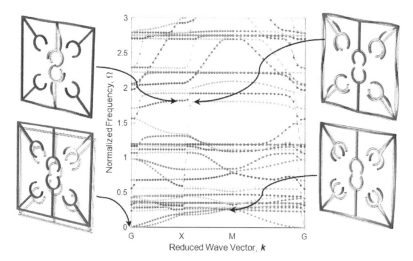

Figure 15. The counterpart of Figure 12 for 70 mm/s.

Figures 13 and 14 reveal that increasing the 4D printing speed from 20 to 40 mm/s does not affect the bandgap range much, changes the stop-band area from 2.01% to 2.07%, and moves the range of bandgap to $\Omega = 1.801 - 1.863$. However, Figures 12c and 15 show an interesting point: after heating–cooling of the diagonal metastructure with the printing speed of 70 mm/s, the whole stop-band area is vanished, and this model allows all the frequencies in the range of 0–3 times of reference frequency to pass. Also, the mode shapes of the structure in some of the high symmetry points and before the bandgap area are depicted in Figure 13. The results presented in Figures 11–15 imply that varying 4D printing speed or heating temperature in the diagonal structure changes the dispersion behaviors significantly and can be manipulated to find an appropriate locally resonant vibration filter. The phenomenon of bandgap switch caused by changing the natural frequency of the active part in the periodic structure is diminished by local resonance changing frequency in different printing speeds.

The parallel metastructure, as shown in Figure 9b displays different dispersion behaviors than the diagonal metastructure (Figure 9a). The dispersion curve for this model is illustrated in Figure 16, where there is no bandgap area, meaning that this structure allows all elastic waves in the frequency range of 0 to 3 to pass. Like diagonal metastructures, the active elements of the parallel structure are also manufactured with three different printing speeds. The configuration of parallel metastructures after the heating–cooling process is depicted in Figure 17, where parts a–c represent the self-bending metastructures 4D-printed at the speeds of 20, 40, and 70 mm/s, respectively. Furthermore, the band structure and mode shapes of parallel structures with self-bending elements of different printing speeds 20, 40, and 70 mm/s after the heating–cooling process are depicted in Figures 18–20, respectively.

As it can be seen in Figures 16 and 18, Figures 19 and 20, the dynamic behaviors of these metastructures are remarkable such that the bandgap area has an increasing–decreasing trend as the structure is heated, revealing self-bending features. Figure 18 shows that the actuated metastructure with self-bending elements 4D-printed at 20 mm/s has a narrow bandgap area in the range of $\Omega = 1.945 - 1.989$ with the amount of 1.48%. However, by using active elements with a higher printing speed of 40 mm/s, the dynamic behaviors change. It is found that the amount of bandgap area increases to 12.32%, and the system exhibits stop-bands in multiple frequencies (Figure 19). These ranges are read as $\Omega = 2.172 - 2.231$, $\Omega = 2.371 - 2.505$, $\Omega = 2.421 - 2.430$, $\Omega = 2.441 - 2.569$, and $\Omega = 2.594 - 2.765$. This implies that this design has a better performance than the others. It can be concluded that this type of 4D-printed architected metastructure has excellent potential in adapting its locally resonant filters from 0 to a significant value such as 12.32%. These bandgaps are generated by Bragg scattering within the medium. In this type of adaptive periodic structure, there is no locally resonant bandgap and the bandgaps are the Bragg type.

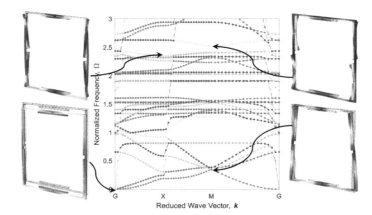

Figure 16. Band structure and mode shapes of the parallel metastructure.

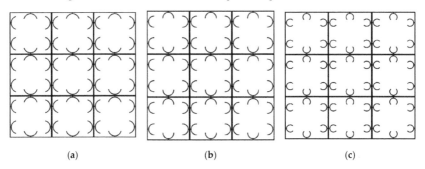

(a)   (b)   (c)

Figure 17. The configuration of adaptive periodic parallel metastructure after heating–cooling process for three different printing speeds: (**a**) 20, (**b**) 40, and (**c**) 70 mm/s.

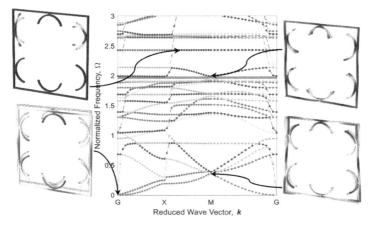

Figure 18. Band structure and mode shapes of the parallel metastructure with active elements printed at 20 mm/s after the heating–cooling process.

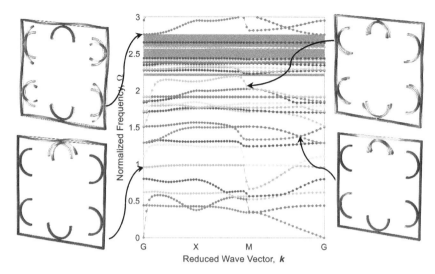

Figure 19. The counterpart of Figure 17 for 40 mm/s.

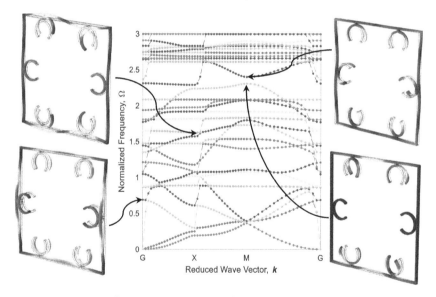

Figure 20. The counterpart of Figure 17 for 70 mm/s.

Finally, the numerical results presented in Figure 20 reveal that by using active elements with a 4D printing speed of 70 mm/s, the bandgap area vanishes. This means that this metastructure, such as the structure before heating–cooling, as shown in Figure 16, propagates all the locally resonant vibration in all directions.

5. Conclusions

This article was aimed at 4D printing adaptive metastructures with locally resonated and Bragg-type stop-bands. The FDM 4D printing technology was implemented to program shape-memory elements during the layer-by-layer deposition process in a functionally graded manner. Experiments were conducted to explore 1D-to-2D self-bending features characterized in terms of 4D printer head speed. Boundary value problems were solved to explain thermomechanical mechanisms behind inducing prestrain during 4D printing and shape recovery after thermal activation. In this respect, a straightforward approach was introduced and implemented into the commercial FE software package of COMSOL Multiphysics, which is much simpler than writing a user-defined material model (UMAT) subroutine or an in-house FE code. The 4D-printed elements were simulated as functionally graded materials whose thermal expansion changed through the thickness direction. The excellent accuracy of the proposed technique was checked via a comparative study with experiments and computational results from the developed in-house FE MATLAB-based solution. Two periodic architected temperature-sensitive metastructures with adaptive dynamical characteristics were conceptually proposed. The COMSOL-based computational tool was then applied to dynamically analyze periodic metastructures with self-bending active elements 4D-printed at different printing speeds. It was found that the metastructures have the capability of controlling elastic wave propagation by forming bandgaps or frequency ranges where the wave cannot propagate. It was observed that the bandgap size and frequency range could be controlled and broadened through local resonances by changing 4D printing speed and thermal excitation. Due to the absence of a similar concept and results in the specialized literature, this article is likely to advance the state-of-the-art tunable metastructures for vibration mitigation and sound attenuation.

Author Contributions: Conceptualization, M.B.; methodology, R.N., M.B. and H.J.; software, R.N. and H.J.; validation, R.N., M.B. and H.J.; investigation, R.N., M.B., H.J., A.Z. and M.F.; formal analysis, R.N., M.B., H.J., A.Z. and M.F.; resources, M.B.; writing—original draft preparation, R.N., M.B. and H.J.; writing—review and editing, R.N., M.B., H.J., A.Z. and M.F.; supervision, M.B.; project administration, M.B. All authors have read and agreed to the published version of the manuscript.

Funding: This research received no external funding.

Conflicts of Interest: The authors declare no conflict of interest.

References

1. Kushner, A.M.; Vossler, J.D.; Williams, G.A.; Guan, Z. A biomimetic modular polymer with tough and adaptive properties. *J. Am. Chem. Soc.* **2009**, *131*, 8766–8768. [CrossRef]
2. Li, Z.; Loh, X.J. Four-dimensional (4D) printing: Applying soft adaptive materials to additive manufacturing. *J. Mol. Eng. Mater.* **2017**, *5*, 1740003. [CrossRef]
3. Bodaghi, M.; Noroozi, R.; Zolfagharian, A.; Fotouhi, M.; Norouzi, S. 4D printing self-morphing structures. *Materials* **2019**, *12*, 1353. [CrossRef]
4. Leng, J.; Lu, H.; Liu, Y.; Huang, W.M.; Du, S. Shape-memory polymers—A class of novel smart materials. *MRS Bull.* **2009**, *34*, 848–855. [CrossRef]
5. Athukoralalage, S.S.; Balu, R.; Dutta, N.K.; Roy Choudhury, N. 3D Bioprinted Nanocellulose-Based Hydrogels for Tissue Engineering Applications: A Brief Review. *Polymers* **2019**, *11*, 898. [CrossRef]
6. Wang, J.; Liu, Y.; Su, S.; Wei, J.; Rahman, S.E.; Ning, F.; Christopher, G.; Cong, W.; Qiu, J. Ultrasensitive Wearable Strain Sensors of 3D Printing Tough and Conductive Hydrogels. *Polymers* **2019**, *11*, 1873. [CrossRef]
7. Soltani, A.; Noroozi, R.; Bodaghi, M.; Zolfagharian, A.; Hedayati, R. 3D Printing On-Water Sports Boards with Bio-Inspired Core Designs. *Polymers* **2020**, *12*, 250. [CrossRef]
8. Keshavarzan, M.; Kadkhodaei, M.; Badrossamay, M.; Ravari, M.K. Investigation on the failure mechanism of triply periodic minimal surface cellular structures fabricated by Vat photopolymerization Additive Manufacturing under compressive loadings. *Mech. Mater.* **2020**, *140*. [CrossRef]
9. Meena, K.; Calius, E.; Singamneni, S. An enhanced square-grid structure for additive manufacturing and improved auxetic responses. *Int. J. Mech. Mater. Des.* **2019**, *15*, 413–426. [CrossRef]
10. Wu, H.; Chen, P.; Yan, C.; Cai, C.; Shi, Y. Four-dimensional printing of a novel acrylate-based shape memory polymer using digital light processing. *Mater. Des.* **2019**, *171*. [CrossRef]
11. Teoh, J.; An, J.; Feng, X.; Zhao, Y.; Chua, C.; Liu, Y. Design and 4D printing of cross-folded origami structures: A preliminary investigation. *Materials* **2018**, *11*, 376. [CrossRef]

12. Spiegel, C.A.; Hippler, M.; Münchinger, A.; Bastmeyer, M.; Barner-Kowollik, C.; Wegener, M.; Blasco, E. 4D Printing at the Microscale. *Adv. Funct. Mater.* **2019**. [CrossRef]
13. Tibbits, S. 4D printing: Multi-material shape change. *Arch. Des.* **2014**, *84*, 116–121. [CrossRef]
14. Sun, L.; Huang, W.M.; Ding, Z.; Zhao, Y.; Wang, C.C.; Purnawali, H.; Tang, C. Stimulus-responsive shape memory materials: A review. *Mater. Des.* **2012**, *33*, 577–640. [CrossRef]
15. Yarali, E.; Baniassadi, M.; Baghani, M. Numerical homogenization of coiled carbon nanotube reinforced shape memory polymer nanocomposites. *Smart Mater. Struct.* **2019**, *28*, 035026. [CrossRef]
16. Shie, M.-Y.; Shen, Y.-F.; Astuti, S.D.; Lee, A.K.-X.; Lin, S.-H.; Dwijaksara, N.L.B.; Chen, Y.-W. Review of Polymeric Materials in 4D Printing Biomedical Applications. *Polymers* **2019**, *11*, 1864. [CrossRef]
17. Bodaghi, M.; Damanpack, A.; Liao, W. Adaptive metamaterials by functionally graded 4D printing. *Mater. Des.* **2017**, *135*, 26–36. [CrossRef]
18. Wang, Q.; Tian, X.; Huang, L.; Li, D.; Malakhov, A.V.; Polilov, A.N. Programmable morphing composites with embedded continuous fibers by 4D printing. *Mater. Des.* **2018**, *155*, 404–413. [CrossRef]
19. Zhang, Q.; Zhang, K.; Hu, G. Smart three-dimensional lightweight structure triggered from a thin composite sheet via 3D printing technique. *Sci. Rep.* **2016**, *6*, 22431. [CrossRef]
20. Zhao, T.; Yu, R.; Li, X.; Cheng, B.; Zhang, Y.; Yang, X.; Zhao, X.; Zhao, Y.; Huang, W. 4D printing of shape memory polyurethane via stereolithography. *Eur. Polym. J.* **2018**, *101*, 120–126. [CrossRef]
21. Zolfagharian, A.; Kaynak, A.; Khoo, S.Y.; Kouzani, A. Pattern-driven 4D printing. *Sens. Actuators A* **2018**, *274*, 231–243. [CrossRef]
22. Bertoldi, K. Harnessing instabilities to design tunable architected cellular materials. *Annu. Rev. Mater. Res.* **2017**, *47*, 51–61. [CrossRef]
23. Schaedler, T.A.; Jacobsen, A.J.; Torrents, A.; Sorensen, A.E.; Lian, J.; Greer, J.R.; Valdevit, L.; Carter, W.B. Ultralight metallic microlattices. *Science* **2011**, *334*, 962–965. [CrossRef]
24. Lee, J.-H.; Wang, L.; Boyce, M.C.; Thomas, E.L. Periodic bicontinuous composites for high specific energy absorption. *Nano Lett.* **2012**, *12*, 4392–4396. [CrossRef]
25. Phani, A.S.; Woodhouse, J.; Fleck, N. Wave propagation in two-dimensional periodic lattices. *J. Acoust. Soc. Am.* **2006**, *119*, 1995–2005. [CrossRef]
26. Matlack, K.H.; Bauhofer, A.; Krödel, S.; Palermo, A.; Daraio, C. Composite 3D-printed metastructures for low-frequency and broadband vibration absorption. *Proc. Natl. Acad. Sci. USA* **2016**, *113*, 8386–8390. [CrossRef]
27. Nanda, A.; Karami, M.A. Tunable bandgaps in a deployable metamaterial. *J. Sound Vib.* **2018**, *424*, 120–136. [CrossRef]
28. Nimmagadda, C.; Matlack, K.H. Thermally tunable band gaps in architected metamaterial structures. *J. Sound Vib.* **2019**, *439*, 29–42. [CrossRef]
29. Baghani, M.; Naghdabadi, R.; Arghavani, J. A semi-analytical study on helical springs made of shape memory polymer. *Smart Mater. Struct.* **2012**, *21*. [CrossRef]
30. Truesdell, C.; Noll, W. The non-linear field theories of mechanics. In *The Non-Linear Field Theories of Mechanics*; Springer: Berlin/Heidelberg, Germany, 2004; pp. 1–579.
31. Bacigalupo, A.; Lepidi, M. Acoustic wave polarization and energy flow in periodic beam lattice materials. *Int. J. Solids Struct.* **2018**, *147*, 183–203. [CrossRef]
32. Liu, J.; Li, L.; Xia, B.; Man, X. Fractal labyrinthine acoustic metamaterial in planar lattices. *Int. J. Solids Struct.* **2018**, *132*, 20–30. [CrossRef]
33. Maurin, F.; Claeys, C.; Deckers, E.; Desmet, W. Probability that a band-gap extremum is located on the irreducible Brillouin-zone contour for the 17 different plane crystallographic lattices. *Int. J. Solids Struct.* **2018**, *135*, 26–36. [CrossRef]

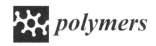

Review

Polymer-Based MEMS Electromagnetic Actuator for Biomedical Application: A Review

Jumril Yunas [1,*], **Budi Mulyanti** [2], **Ida Hamidah** [2], **Muzalifah Mohd Said** [3], **Roer Eka Pawinanto** [4], **Wan Amar Fikri Wan Ali** [1], **Ayub Subandi** [1], **Azrul Azlan Hamzah** [1], **Rhonira Latif** [1] and **Burhanuddin Yeop Majlis** [1]

[1] Institute of Microengineering and Nanoelectronics, Universiti Kebangsaan Malaysia, Bangi 43600, Selangor, Malaysia; p86814@siswa.ukm.edu.my (W.A.F.W.A.); p93203@siswa.ukm.edu.my (A.S.); azlanhamzah@ukm.edu.my (A.A.H.); rhonira@ukm.edu.my (R.L.); burhan@ukm.edu.my (B.Y.M.)

[2] Faculty of Engineering and Vocational Education, Universitas Pendidikan Indonesia, Jl. Dr. Setiabudhi 207, Bandung 40154, Indonesia; bmulyanti@upi.edu (B.M.); idahamidah@upi.edu (I.H.)

[3] Faculty of Electronics and Computer Engineering (FKEKK), Universiti Teknikal Malaysia Melaka (UTeM), Hang Tuah Jaya, Durian Tunggal 76100, Melaka, Malaysia; muzalifah@utem.edu.my

[4] Malaysia-Japan International Institute of Technology (MJIIT), Universiti Teknologi Malaysia (UTM), Kuala Lumpur 54100, Malaysia; roer.eka@gmail.com

* Correspondence: jumrilyunas@ukm.edu.my; Tel.: +603-8911-8541

Received: 30 March 2020; Accepted: 28 April 2020; Published: 22 May 2020

Abstract: In this study, we present a comprehensive review of polymer-based microelectromechanical systems (MEMS) electromagnetic (EM) actuators and their implementation in the biomedical engineering field. The purpose of this review is to provide a comprehensive summary on the latest development of electromagnetically driven microactuators for biomedical application that is focused on the movable structure development made of polymers. The discussion does not only focus on the polymeric material part itself, but also covers the basic mechanism of the mechanical actuation, the state of the art of the membrane development and its application. In this review, a clear description about the scheme used to drive the micro-actuators, the concept of mechanical deformation of the movable magnetic membrane and its interaction with actuator system are described in detail. Some comparisons are made to scrutinize the advantages and disadvantages of electromagnetic MEMS actuator performance. The previous studies and explanations on the technology used to fabricate the polymer-based membrane component of the electromagnetically driven microactuators system are presented. The study on the materials and the synthesis method implemented during the fabrication process for the development of the actuators are also briefly described in this review. Furthermore, potential applications of polymer-based MEMS EM actuators in the biomedical field are also described. It is concluded that much progress has been made in the material development of the actuator. The technology trend has moved from the use of bulk magnetic material to using magnetic polymer composites. The future benefits of these compact flexible material employments will offer a wide range of potential implementation of polymer composites in wearable and portable biomedical device applications.

Keywords: polymer composites; microelectromechanical system (MEMS); electromagnetic (EM) actuator; magnetic membrane; microfluidic; biomedical

---

1. Introduction

Over the past few years, there has been an increasing demand on the employment of flexible materials for various applications in biomedical field. This has led to the significant growth of the movable structure development [1,2]. The flexible material having good mechanical properties with high surface strength and high elasticity has enabled tremendous innovation in the development of microelectromechanical systems (MEMS) devices in which the electrical and mechanical property of the material are the most important characteristics of the technology [3]. One of the most interesting materials is polymer that currently can be found in various biomedical instrumentation due to its excellent mechanical properties, compactness, precise control and biocompatibility as well [4].

The flexibility characteristic of polymer is beneficial in obtaining large and controlled structure deformation of the movable parts. These movable parts include diaphragm (thin membrane), pillars, cantilevers or the combination of pillars and movable structures [5,6]. This class of functional material plays very important role in the development of MEMS electromagnetic (EM) actuators, for example the microfluidic delivery system found in drug delivery, bio-cell preparation system and lab on chip [7]. The system can also include micropumps, microvalve, micromixer, microgripper and micromanipulators [8–11].

Studies on electromagnetically driven MEMS actuators in the field of biomedical instrumentation are currently increasingly popular in which the improvements of the mechanical structures and the material properties of the movable part became the most interesting topics. The development studies were done in order to enable efficient and precise structure movement for control, manipulation or analysis purpose of the biomedical samples [12,13]. These studies also have led to the invention of flexible structure possessing sensitive interaction with magnetic induction, to be the most important mechanism in electromagnetic actuation. The moving structures should be made of soft and elastic material, able to continuously vibrate and capable of reacting to mechanical pressure and magnetic field exposures [14].

Several reports have been recently published to introduce the interaction between magnetic flux generated from electromagnetic coil and rotating magnet field [15,16]. This interaction is the basic principal operation of the electromagnetic actuator that produces magnetic force to enable the movement of a movable structure. The basic electromagnetic actuator structure consists of a flexible movable membrane, electromagnetic coil, magnetic chamber or spacer and bulk permanent magnet. Initially, a thin membrane attached with permanent magnet has been the common structure used as the moving membrane of the MEMS electromagnetic actuator [17]. Unfortunately, the structure with attached bulk magnet suffers from high volume and low reliability, especially when the membrane operates in long vibration mode [18]. Therefore, some innovations in the material structure have been developed in order to obtain a compact and reliable actuator.

The MEMS structures are usually made of glass, silicon, silicon nitride and metals [19,20]. Those materials are the common materials in MEMS technology due to the excellent mechanical properties and matured technology process [21]. However, silicon and glass are easy to break as they have low fracture strain which is about 0.1% [22–24]. Meanwhile, metals are very sensitive to chemical and environmental effect [25]. Some other disadvantages of those conventional MEMS materials, especially for the use as movable structure, are fragile and low flexibility. These drawbacks make them less favorable compared to polymers.

On the other hand, polymers in MEMS have been used since several years ago as a photosensitive material [26], sacrificial layer [27], passive structure for microchannel [28], microchamber and passive micromixer [29] and as the functional layer of micro-structured devices, such as actuators [30] and sensors [31,32]. Polymer has good mechanical properties with Young's modulus lower than silicon and metal, which makes it highly elastic and at the same time possesses high strength [33–38]. In conjunction with MEMS actuators, the mentioned mechanical properties are useful in obtaining large membrane deformation under external magnetic stimulus. Furthermore, the most important fact is that the polymeric structure of MEMS device can be fabricated in inexpensive way, cheaper than silicon-based micro-processing cost [39–41].

It was also reported that microstructures working under extreme vibration condition like actuators need enhancement in terms of material quality, design and technological concepts in order to increase the lifetime and effectiveness of the structures [38]. Therefore, some magnetic polymers become more preferable as the structures will have high elasticity, easy to fabricate and photo-patternable.

Some popular polymers have been identified and explored to become the flexible material for actuation purposes. The common actuator materials that have been reported in the literatures include PMMA, parylene, polyimide and PDMS elastomer. The properties of those materials are summarized in Table 1.

Table 1. Material properties of popular polymers used in microelectromechanical systems (MEMS).

Polymer Name	Density	Young's Modulus (GPa)	Poisson's Ratio	Thermal Expansion Coefficient @25 °C (10^{-6} K^{-1})	Thermal Conductivity (W/mK)	Property Utilized	Process
PMMA [41,42]	1.17–1.2	3.1–3.3	0.35	70–90	0.186	Little elasticity, optical property	LIGA, Hot embossing
Parylene [43]	1.289	4.5	0.4	35	-	Vapor barrier	Coating
PDMS [29,39,44]	0.97	0.36–0.87	0.5	310	0.18	Elasticity	Molding
Polyimide [45–47]	1.42	3	0.34	30–60	0.1–0.35	Little elasticity	Coating

2. MEMS Actuators

In general, MEMS actuators can be driven either by mechanical actuation or non-mechanical actuation. Mechanical actuation mechanism with a diaphragm (membrane) as the moving part is primarily utilized in MEMS devices [48]. Compared to the non-mechanical actuator, the mechanical actuator has many advantages in terms of controllability, high vibration rate and large membrane deformation [49].

A large number of mechanical microactuator devices has been demonstrated including microrelays, microvalves, optical switches and mirrors, micropumps and many others that can be found in various applications. These actuators use different mechanical actuation principle such as piezoelectric, electrostatic, electromagnetic, thermo-mechanic, thermo-pneumatic and shape memory. Table 2 shows a comprehensive analysis of MEMS mechanical actuators, describing different types of energy exchange mechanism used to obtain kinetic movement, the devices' structure, the advantages and the typical applications of each MEMS mechanical actuator.

Table 2. Typical MEMS mechanical actuator devices, structure and their working principle.

Working Principle	Schematic of Actuator System	Advantages	Disadvantages	Typical Applications
Piezoelectric [9,50,51]		High pressure Fast response	Complicated process High input voltage, low reliability	Micropump, microvalve, microgripper
Electrostatic [52–55]		Low Power Fast response Controlled large deformation through input voltage	Small membrane deformation, low reliability	Micromotor, microshutter, micromirror microrelay, micropump
Electromagnetic [9,12,49,56]		**High pressure.** **High membrane deformation Easy control thrugh input current Fast response Large frequency range.**	**Large size.** **Thermal effect**	**Micromotors, micro relay, switch, micro pump, valve, mixers, microspeaker and magnetostrictive**
Polymer composite Electroactive [57]		High deformation Low power Ability to work at wet environment Low footprint	New actuation mechanism. Complicated structure and process Very limited application	Micro robotic, micromanipulators
Thermo-pneumatic [11,58]		High pressure	Specific material high power consumption long response time. Limited application	Micropump, microvalve, inkjet printhead

The major advantages of electromagnetic actuation are the generated high magnetic force that enables large membrane deflection and high tunable frequency capability. In addition, rapid generation of electromagnetic field enables membrane deformation in 2 directions with very fast vibration rate [10]. Additionally, an electromagnetic actuator is capable of precisely tuning the input power. The power consumption in EM actuators between 13 mW to 7 W is the widest range among the other types of actuators [59]. However, high power dissipation and large area consumption could be the drawbacks of the system.

Not many designs for magnetic microactuators specifically used in biomedical field are reported in literature. Table 3 shows the developed magnetic microactuator devices for biomedical application. The application of these actuators are classified into biosample delivery/transport, biosample preparation and biocell manipulation. Mainly, the actuator functioned as a microfluidic handling system for samples delivery in a drug delivery system and lab on chip. There is a high interest from industry in the implementation of the electromagnetically driven microactuators for a broad range of biomedical applications.

Table 3. Common magnetic actuator devices used for biomedical applications.

References	Actuating Element (Structure, Material or Method)	Magnet Type	Input	Specifications
Biosample Delivery and Transport				
Yamahata et al. 2015 [60]	PDMS membrane & magnet	Iron powder	33–150 mA	Flowrate: 0.4–1.6 mL/min Frequency: 6–12 Hz
Büttgenbach, 2014 [61]	EM Micromotor rotation & polymer magnet	90 wt% ceramic ferrites + polymer	70 mA	Forces: 1.2 mN Torque: 10 µNm
Lee et al. 2011 [62]	Silicon catheter	Electroplated nickel	70 to 1500 Hz (resonant frequency)	Angle > 60°
Zhou & Amirouche, 2011 [63]	PDMS membrane & magnet	NdFeB or CoNiMnP plate	90–180 mA	Magnetic Force: 16 µN Flowrate: 319.6 µL Frequency: 36.9 Hz
Biosample Preparation				
Nouri et al. 2017 [64]	Magnetohydrodynamic interaction with permanent magnet	Fe_3O_4 nanoparticles	3000 Gauss	Mixing time: 80 s Mixing index: 0.9 s
Liu et al. 2016 [65]	PDMS with permanent magnet	Magnetic composite (carbonyl iron)	6 V, 18 Hz	Mixing time: 2 min Flow rate: 20 µL/s
Biocell and Drug Particles Manipulation				
Banis et al. 2020 [66]	water-soluble ferrofluid material (FluidMAG lipid)	Electromagnetic coils	4 to 8 A Magnetic particle size 100 nm	Droplet velocity 135 µm/s
Rinklin et al. 2016 [67]	Magnetophoretic attraction of microbeads	carboxyl functionalized particles (Dynabeads) and laminated magnetic NiFe parts	5, 10 and 15 mA	Maximum particle levitation height of approximately 10 µm
Chen et al. 2015 [68]	PDMS tweezer with hexapole yoke	10 layers of laminated magnetic NiFe parts	feedback control at a speed of up to 1 kHz	Maximum force = 400 pN, force distribution with actuation from −30 µm to 30 µm
Choi et al. 2000 [69]	silicon cantilever	Encapsulated permalloy	N/A	N/A

Electromagnetic Actuators Principle

The basic mechanism of electromagnetic actuation involves the interaction between magnet and electromagnetic field that intensively generates the magnetic force. This interaction produces high frequency vibration of the movable structures, such as membrane and pillars, hence enables various implementation of biomedical instrumentation.

Thielicke et al., [70] explained that the actuation principle depends on structural dimension, response time, torque, max power consumption, the technology used and the applied forces. The forces are classified into 2 main groups, namely external and internal forces. Electromagnetic actuators fall in external forces category as the forces are produced from the magnetic fields interaction occurred in the gap between the stationary and moving parts.

In general, the magnetic membrane actuation is achieved by the deformation of the movable membrane due to the generated magnetic force acting onto the membrane. The common structure of a magnetic actuator is schematically displayed in Figure 1a. The system consists of the magnetic field generator part (electromagnetic coil) and the magnetic membrane part (flexible membrane plus an attached magnet) [71,72].

Through the interaction between magnet and electromagnetic coil, a vertical magnetic force acting on the magnetic membrane with vertical magnetization on z-axis is generated. The magnetic force known as Lorentz force F_{mag} is given by the following integral over the volume V of the body [73]:

$$F_{mag} = M_z \int_v \frac{\partial H_z}{\partial z} dV \tag{1}$$

where, $Fmag$ is the magnetic force acting on the magnetic membrane, M_z is the magnetic induction from the permanent magnet, $\frac{\partial H_z}{\partial z}$ is the magnetic field gradient generated by the electromagnetic coils and dV is the volume of the permanent magnet. The correlation between magnetic force applied on to the membrane and the resulting membrane deformation h_z can be derived from the Equation (1):

$$h_z = C \frac{Fmag\ l_m}{D} \tag{2}$$

where, l_m is the membrane size, C is a constant depending on the shape and geometry and D is the material characteristics of the membrane that is defined by:

$$D = \frac{E\ t_m^2}{12\ (1 - v^2)} \tag{3}$$

where E is the Young's Modulus, v is the Poisson's ratio while, t_m is the thickness of the membrane.

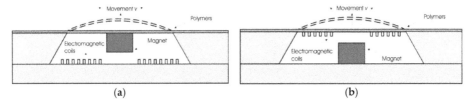

(a) (b)

Figure 1. Cross sectional view of an initial electromagnetic (EM) actuator, (**a**) with magnetic membrane-based moving parts [50], (**b**) with embedded planar coil-based moving parts [74].

Another approach to introduce the principal of the actuation mechanism has been described by Pawinanto et al. [71] and Sugandi et al. [74]. Here, planar electromagnetic coil wires are embedded inside or attached on the movable membrane surface, as shown in Figure 1b. When an electrical current is supplied to the planar coil wires, a magnetic flux induction from the permanent magnet onto the wires is achieved. Through this induction, the magnetic force $Fmag$ is generated and acting onto the membrane that finally causes the periodical actuation of the membrane structure.

At the location of the coil, magnetic field makes an angle θ with the normal surface (vertical axis) and the magnetic force ($Fmag$) between a current carrying wire and a permanent magnet can be expressed as given by [22]:

$$\vec{F_{mag}} = \sum_{i=1}^{N} 2\pi R_i I \times \vec{B_r}(R_i) \times \sin \theta \tag{4}$$

where I is the coil current, R_i the radius of each turn coil, $\vec{B_r}$ the radial component of magnetic field in the coil plane and θ is angle direction of magnetic field to vertical axis. Therefore, total force for a single turn coil is given by:

$$\vec{F_{mag}} = I \, (l \times \vec{B_r} \times \sin \theta) \tag{5}$$

with l represents the total length of a single turn coil with a radius r. Using both equations, we can see where the force vector direction acted. The induced electromagnetic force is principally based on the magnetic interaction between the current carrying coils, permanent magnets and flexible membrane materials [71]. It works vice versa, either the membrane with embedded wire moves or the magnet moves.

3. MEMS Fabrication of Polymer-Based Actuator

3.1. Fabrication of EM Actuator

There are several mechanical actuation mechanisms related to the function of the membrane such as vibration, peristaltic and flexural plate wave [75]. Some actuators are constructed with flat movable membranes [45], some others are equipped with pillars or cilia, as found in micromixers [76]. For these purposes, certain MEMS fabrication methods with high resolution pattern are needed in order to create three-dimensional structures on the membrane. It should be noted that the fabricated membrane structure must be flexible enough to generate movement and able to withstand the pressure acting onto the surface. The patterned structures on the membrane were also predicted to improve the membrane's flexibility.

The common method used in fabricating a polymer membrane with three-dimensional (3-D) structure is soft lithography or micro-molding. Soft lithography technique for polymer-based MEMS device was introduced in 1990 by Varadan [77]. Among the advantages of soft lithography techniques compared to conventional optical lithography techniques are the unlimited machining resolution of the emission and dispersion of optical waves and the turbulence in the resin. In addition, soft lithography with elastomer sealants has the advantage of precise pattern on the target surface and easy to remove from the mold. All of these advantages make soft lithography a great attractive and highly potential technique to be used in the field of microfabrication process [78].

Most of the polymer membranes fabricated through soft lithography technique do not have their own mechanism in order to function as an actuator. They need an external stimulation either from a permanent magnet or an electromagnet. Via this concept, an actuator disc, a magnetized permalloy strip, a bulk magnet or an embedded electromagnetic coil can be integrated into the polymer membrane structure to generate force for the membrane deformation purpose [79–82]. Soft lithography technique is not only an inexpensive and simple fabrication process but it can also manipulate the texture of the polymer membrane during fabrication to control its flexibility which is vital for membrane actuation [83–85].

Some examples of soft lithography process in the fabrication of polymer-based MEMS structure were reported by Ghanbari et al. [86] and Yunas et al. [87,88]. The microactuator part can be fabricated separately. Thus, the fabrication process can start with the electromagnetic part (1), followed by the fabrication of magneto-mechanic part (2) and finally with the bonding of both parts using epoxy (3). The detailed fabrication process of a micropump system is shown in Figure 2. The electromagnetic coil pattern is first created followed by the deposition of planar copper (Cu) microcoil wire (a) and (b). The coil structure is formed after the lift-off process (c) and (d). Next is the fabrication of the magneto-mechanical part that involves the patterning of mold master using SU8-based photolithography process (e). Then, the polymer membrane is fabricated by pouring the PDMS onto the pre-patterned structure (f) followed by peeling-off of the material (g) before transferring it onto a spacer surface. The final step of the process is the attachment of the permanent magnet onto the transferred membrane and all fabricated parts are bonded together using epoxy glue (h).

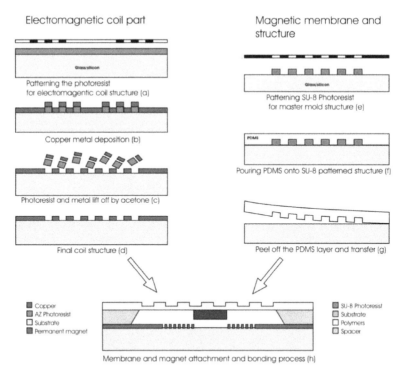

Figure 2. Schematic process step for the fabrication of polymer-based MEMS EM actuators with micro-pillar structures using the soft lithography process technique.

Another approach to create membrane with 3-D (three dimensional) structures has been reported by Xu and Cui [89]. They used hot embossing technique to fabricate an actuator membrane by constructing the membrane layer-by-layer (LbL). In the process, silicon molds were fabricated using a conventional UV lithography and wet, etching technique. The hot press technique was then used to transfer the design structure from silicon molds to PMMA sheets. The hot press molding technique involves the simultaneous application of heat and pressure in the fabrication of a polymeric membrane.

Furthermore, 3-D structures can be created using 3-D printing technique that can print biocompatible polymers or devices at required dimensions based on the printer's resolution. The technique offers more complex and sophisticated design that can be realized at micro-scale which could not be done with conventional method like soft lithography [90,91]. There have been also several studies reporting the usage of 3-D printing to fabricate a part of MEMS device such as the stereolithographic (SL) 3-D printer that fabricates a thin membrane from poly(ethylene diacrylate) resin [92]. The membrane was then pneumatically pressed to get the thickness smaller than 25 μm.

Zhou et al. [93] also reported in 2019, that a polymer actuator membrane with a thickness of 100 μm was successfully fabricated using a 3-D multijet printer (MJP). The printed membrane was able to deform in order to close and open the microchannel and fully functioning as a valve. Another novel 3D-printed electromagnetically driven fluidic valve was fabricated by projection–stereolithography (PSL) in combination with functional elements such as the permanent magnets [94]. There was also a study on the fabrication of a whole MEMS device using 3-D printing technique that met minimum requirement for biocompatible standard [95].

3.2. Fabrication of Magnetic Polymer Composites Membrane

Embedded magnetic particles in polymer would be the future functional material for many types of biomedical devices. It becomes a new composite material that possesses the flexible mechanical characteristic and exceptional magnetic responsive features [96]. The implementation of magnetic polymer composite as the material structure for the actuator membrane could overcome the need of a bulk permanent magnet. The soft and flexible properties of polymeric membrane would tend to rupture and break when a bulk structure is placed on it, like the bulky permanent magnet attached onto an actuator membrane of a micropump [88].

One of the methods in fabricating magnetic polymer composite MEMS membrane is through a synthesis method using mechanical stirring under sonication in which a PDMS-based polymer was mixed with NdFeB magnetic particles having the size ranging from 50 to 100 um [9]. The deformation capability of the membrane has been tested, by which the highest deflection of 9.16 μm at 6 vol% magnetic particles density has been measured with an applied magnetic field density of only 0.98 mT.

Here, the PDMS is considered as the most popular material for flexible biomedical device applications. Apart from its biocompatible property, the mechanical properties can be manipulated via controlled ratio of the polymer base and curing agent [97]. The PDMS-based membrane has been successfully fabricated with the integration of the magnetic particles from 2% up to 30% distribution across the membrane. The magnetic membrane can be deflected when its magnetic field interacts with the magnetic flux formed from the current flow in the coils. The fluctuating movement of the membrane is governed by the applied current of only several milliamperes.

Recently, Tahmasebipour and Paknahad have fabricated nano-magnetic membrane made of PDMS–Fe_3O_4 for the application of valveless electromagnetic micropump [98]. Nano sized particles of Fe_3O_4 were mixed within the PDMS layer in order to create the magnetic membrane. The composite magnetic membrane is compatible with living tissues and has great magnetic stability. The embedding of nanoparticles in polymer however can cause agglomeration problem due to the attractive forces between the particles. Therefore, different approaches have been proposed to minimize particle agglomeration, such as particle encapsulation with polymeric material [99] or ceramic coating [100] or by implementation of surfactant [101].

4. Application of Polymers for Electromagnetic Actuators

4.1. Magnetic Polymer Composite-Based Microactuators

A flexible membrane with embedded magnetic particles having small particle size would have many advantages, because the magnetization and the magnetic anisotropy of the particles can be much greater than a bulk magnetic specimen [102]. The magnetic polymer composite is very light, hence would not significantly affect the mechanical properties of the polymer. Hence, this magnetic polymer composite membrane enables actuators to have larger deflection with a controllable actuation forces, compared to silicon or metal-based actuators [103]. On other hand, with the help of photo sensitive mold master material, the polymer composite would be able to be patterned and transferred onto the substrate as suspended movable part and other MEMS passive structures as well. Thus, the material composite can find its potential application as sensitive actuator for fluid injection, valves, magnetic recording media, mechanical relays, optical mirror and switch and other mechanically moving part driven by magnetic fields [15,104–106].

The evolution of magnetic material used for the actuator membrane shows a transition from bulk magnet to matrix magnet and now to magnetic polymer composite. The research of magnetism in electromagnetic actuator has then been extended by reducing the size of the magnetic particles embedded in the polymer membrane from micro to nanometer. Here, the evolution of the magnetic membrane is described in Figure 3.

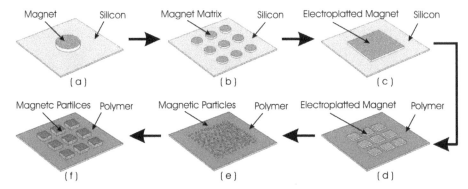

Figure 3. Development history of magnetic polymer composite-based MEMS actuator membranes, (**a**) silicon-based membrane with attached permanent magnet [71], (**b**) silicon membrane with attached small permanent magnet in matrix form [19], (**c**) silicon membrane with electroplated magnetic material [19], (**d**) polymer membrane with electroplated magnetic material [107,108], (**e**) polymer membrane with embedded magnetic particles [9], (**f**) polymer membrane with three-dimensional matrix structured embedded magnetic particles [88].

Initially, silicon material was used as the membrane, which was bonded with a single bulky permanent magnet glued on the top of the membrane. Then, smaller permanent magnet with matrix structure was used to replace the single magnetic bulk in order to reduce the membrane stiffness. In 2002, the use of polymer material as the actuator membrane had been started and the permanent magnetic layer was created on the polymer membrane via electroplating [107]. The concept was then extended with the use of arrays of electroplated permanent magnetic layer [108]. Finally, the electroplated permanent magnet has then been replaced with the embedded magnetic particles, producing a magnetic polymer composite membrane with significant improved performances [88].

The current status of the magnetic polymer composite membrane for biomedical application was reported by Said et al. [88]. They developed a matrix patterned magnetic polymer composite for actuator membrane that is integrated with the micropump for bio-sample injection. The composite membrane is made of polydimethylsiloxane (PDMS) mixed with NdFeB magnetic particles and patterned into blocks of matrix.

To this concept, the magnetic composite actuator membrane containing 6% NdFeB was capable of generating a maximum membrane deflection up to 12.87 μm [9]. As shown in Figure 4, the magnetic property of NdFeB polymer composite is strongly related to the amount of magnetic particles embedded in the polymer. Thicker polymer layer with more NdFeB particles produces larger magnetization. However it doesn't affect the change in coercivity. A 139 μm membrane thickness shows a saturated remanence magnet of 37.637 mT.

Some other potential applications of magnetic polymer composite in sensors and actuators were reported by Samaniego et al. [109]. They studied the resultant of magnetic polymer composite to fabricate soft robots by squeegee–coating method. The soft robots have flexible and compliant bodies resulting in higher degrees of freedom and improved adaptability to their surroundings. Therefore, the robot can be used for minimal invasive surgery (MIS) in order to reduce patients' trauma, pain and recovery time [110]. The soft polymer-based magnetic actuator was fabricated by mixing ferromagnetic microparticles (PrFeB) with polymers precursor before its curing. The soft robots were magnetized under 1 T of uniform magnetic field.

The magnetic polymer composite can also be integrated to the artificial cilia in a microfluidic system. Zhang et al. demonstrated the versatile microfluidic flow generated by molded Magnetic Artificial Cilia (MAC) [111]. The MAC can cause versatile flows by changing the magnetic actuation mode. This on chip microfluidic transport does not require tubing or electrical connections, reducing the consumption of reagents by minimizing the "dead volumes", avoiding undesirable electrical effects and accommodating a wide range of different fluids.

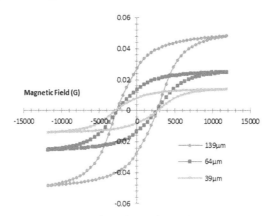

Figure 4. Hysteresis loop of 6% NdFeB polymer composite.

4.2. Polymer-Based Electromagnetic Actuators for Micropumps and Microvalves

Most important properties of polymer in micropump system as the movable membrane of the actuator are its flexibility and high surface strength. In general, microfluidic systems are made up of input and output tubes, channels and pump chambers. Beside the membrane, the valves are also important in ensuring fluid flow direction and to regulate the flow rate. Some other micropumps are designed valveless that improve the reliability of the system and reduces the clogging effect [112].

Wang et al. [113] reported a micropump comprising a magnetic PDMS diaphragm, a planar microcoil and microfluidic channel. When an electric current is applied to the microcoil, an electromagnetic force is generated at the magnetic diaphragm. The deflection of PDMS diaphragm generate a push–pull action of the membrane hence creating pressure difference within the chamber and microchannel and subsequently causing the fluid flow. Their EM micropump achieved a maximum pumping rate of 52.8 µL/min with diaphragm displacement of 31.5 µm induced by a microcoil input current of 0.5 A.

EM micropump using PDMS encapsulation layer mounted with small permanent magnets was reported by Pan et al. [114]. The membrane of the pump actuator which is driven by magnetic motor shaft or microcoil used two one-way check valve using a micropipette and heat sink tubing. The magnetic motor shaft was a small DC motor (6 mm in diameter and 15 mm in length) with a neodymium–iron–boron permanent magnet embedded in its shaft. The EM micropump achieved a maximum pumping rate of 774 µL/min with magnetic motor shaft and 1 mL/min with microcoil driven pump. Microcoil driven pump has shown higher flowrate and much higher power consumption.

Furthermore, an EM micropump with embedded planar coil in the thin PDMS membrane reported by Yin et al. (Figure 5) [115]. The size of the membrane is 7 mm in diameter and achieved 50 µm deflection with an applied current of 500 mA. The resonant frequency is about 1.43 kHz. Fluids in the microfluidic chip were driven forward by a local pneumatic pressure provided by the membrane. This EM pump was claimed to have a pumping volume flow rate of 2 µL/min.

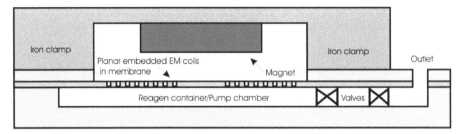

Figure 5. Polydimethylsiloxane (PDMS)-based EM micropump and valves with embedded planar microcoil.

A valveless EM micropump reported by Yamahata et al. [60] used composite magnets replacing the bulk permanent magnets (Figure 6). The composites magnet was developed using PDMS polymer mixed with 40% of NdFeB powder with a mean size of 200 μm. The actuator was driven by a 1500 turns coil supplied with sinusoidal current of 150 mA with soft magnetic core in the center to strengthen the magnetic field. The actuator membrane could deform up to 0.25 mm and pumping water at the flowrate of 400 μL/min with resonant frequency of 12 Hz by applying nozzle/diffuser microfluidic system. The passive structure of the pump system is made of four PMMA layers consisting of capping layer, channel and chamber layer and also spacing layer.

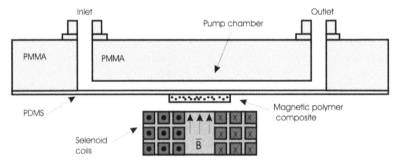

Figure 6. Schematic of a valveless EM micropump involving PDMS and PMMA materials and utilizing magnetic composite membrane to replace the bulk permanent magnet.

Another valveless EM micropump having a composite magnetic membrane of Fe + PDMS was reported by Nagel et al. [116]. Weight concentration varying from 25%–75% of iron particles with the size of below 10 μm were mixed in PDMS. Nickel coated NdFeB magnet was used to interact with the actuator membrane and moved up/down by a crankshaft. This micropump used valveless microfluidic system with a 6-mm diffuser/nozzle that has produced a maximum pumping flowrate of 35 μL/min at a frequency of 1.67 Hz.

Shen and Liu fabricated a PDMS-based magnetic composite membrane with IPDP thin film stacked design (Figure 7a) and embedded system (Figure 7b) [117]). An iron-particle-dispersed PDMS (IPDP) was a mixture of iron particles with the average size of 55 μm, PDMS and its curing agent. The mixing ratio of IPDP was 10:10:1. The micropump used a valveless type microfluidic system and electromagnet block which connected to power supply and combiflex. Micropump with IPDP embedded design had a pumping flowrate of 1.623 mL/min at a frequency of 6–7 Hz and 30 V of supply voltage which was higher compared to the flowrate of stack design.

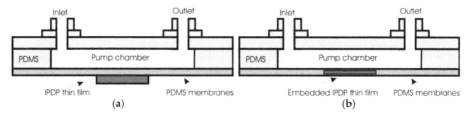

Figure 7. Schematic of iron particle dispersed PDMS (IPDP) for valveless micropumps, (**a**) stacked thin film design, (**b**) embedded thin film design.

A complete PDMS-based micropump including the structure, the membrane and the valveless microfluidic system was reported by Zhou and Amirouche [63]. The actuator membrane used a thin NdFeB magnet encapsulated at the center of the PDMS. Maximum deflection of the actuator membrane was 36.36 μm. DI water is pumped with maximum flow rate of 319.6 μL/min at a frequency of 36.9 Hz with supply current of 0.18 A.

Said et al. [118] combined the bulky permanent magnet with magnetic composite membrane to improve the reliability of the membrane and at the same time to increase the magnetic induction. The hybrid structure could produce a magnetic flux density of 37.637 mT enabling a controllable peristaltic pumping of microfluidic sample with a flowrate of 6.6 μL/min. When the bulk permanent magnet was removed from the micropump system and only left with the flat membrane composite alone, the micropump produced a very slow flowrate of 6.52 nL/min. Hence, the micropump could deliver a very precise dose for drug delivery system.

Electromagnetically actuation of flexible membrane incorporating microvalve for micropump application has been also reported by Sadler et al., as shown in Figure 8 (left) in a closed mode and (right) in an open mode [119]. The normally closed magnetic microvalve has both fluidic and electrical connections bonded to a glass motherboard. The microvalve comprised a magnetoactive membrane, a stationary valve seat and inlet/outlet channel. The magnetoactive membrane as a diaphragm plated with permalloy films on the top will interact with electromagnet flux generated by inductor to produce the magnetic force. A polymer film was attached to the system to ensure there is no leak. The force would lift up the membrane from the valve seat thus opened the valve and allowed the fluids to flow from inlet to outlet due to pressure difference.

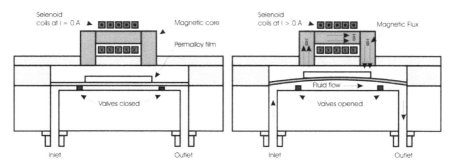

Figure 8. Electromagnetic actuator incorporating magnetic valves, (**left**) closed mode, (**right**) open mode.

Polymer-based microvalves used to manipulate the fluid flow have been reported by Gaspar et al. [120]. The actuation of the valve is based on the principle that flexible polymer walls of a liquid channel can be pressed together by the aid of a permanent magnet and a small metal bar. In the presence of a small NdFeB magnet lying below the channel of interest, the metal bar is pulled downward simultaneously pushing the thin layer of PDMS down thereby closing the channel and stopping any flow of fluid. Furthermore, Bute et al. [121] reported that the flow manipulation and proper operation of the valve depends on thickness and percentage load of magnetic material in the membrane as well as dimensions of channel, chamber and membrane with respect to the location of outlet channels, while Nakahara et al. [122] reported the use of photosensitive polymer composites for the fabrication of magnetically driven microvalve arrays in a μTAS (μ- total analysis system)

To summarize, since 1995 there have been many developments in microactuator device design and materials for micropumps and microvalves incorporating elastic membrane, as listed in Table 4. The magnetic polymer-based micropumps are working with various applied frequency and various reading fluid flow rates ranging from 6.5 nL/min to 6.8 mL/min have been obtained. Obviously it is found that after the use of silicon, polymers have become the subject of researcher's choice to build the actuator membrane for micropump. Since 2005, magnetic polymers composites have become the promising material to replace the conventional bulky permanent magnet.

Table 4. Development of polymer-based MEMS electromagnetic actuators for microfluidic pump applications.

Year	Membrane Structure	Flowrate	Frequency	References
1995	Thermoplastic molding bulk permanent magnet	780 μL/min	5 Hz	Dario et al. [123]
1999	Silicon rubber	2.1 mL/min	50 Hz	Bohm et al. [124]
2000	PDMS + plate alloy	1.2 μL/min	2.9 Hz	Khoo dan Liu [125]
2005	PDMS + bulk permanent magnet	774 μL/min	n.a	T.Pan et al. [114]
2005	PMMA and composite PDMS + powder NdFeB	400 μL/min	12 Hz	Yamahata et al. [60]
2006	Composite PDMS + powder Fe	35 μL/min	1.73 Hz	Nagel [116]
2007	PDMS + bulk permanent magnet	2 μL/min	n.a	Yin et al. [115]
2008	PDMS + bulk magnet NdFeB and PMMA	6.8 mL/min	20 Hz	M.Shen et al. [126]
2010	Composite PDMS + powder Fe	1.623 mL/min	6–7 Hz	Shen and Liu [117]
2011	Composite PDMS + plated NdFeB	319.6 μL/min	36.9 Hz	Zhou and Amirouche [63]
2015	PDMS + magnet pad	n.a	28–30 Hz	Dich et al. [127]
2017	Composite PDMS + NdFeB particles	6.52 nL/min	1 Hz	Said et al. [88,118]

It can be concluded that the polymer-based micropump and valves were able to precisely control the delivery of the fluidic sample and obtained fluid flow range from 10 mL/min down to several nL/min. Innovations in the membrane material and structure and the use of the latest technology in several ways are still necessary to meet the needs and requirements of the biomedical instrumentations.

4.3. Polymer-Based Active Micromixer

Magnetic polymers found its important role in microfluidic mixer system which is mostly used as the basic material for the passive part of the system such as the channel and chamber formation in lab on chip system. On the other hand, the polymer has been playing the potential role as an active part in order to improve the mixing performance of the microfluidic system, especially for the bio-cell analysis. This is called active microfluidic mixer, which means that the mixing mechanism is due to the turbulences of the fluidic sample inside the mixer chamber, usually based on magneto-hydrodynamic and magnetic structure actuation, which is driven by an external magnetic field [128].

In 2018, Tang et al. presented a research on embedded flexible magnets in PDMS membrane [129]. Three designs were introduced and compared, namely (a) concentric type with the magnetic material in the center of the membrane; (b) eccentric type with the magnetic material offset from the center of the membrane and (c) split type with two regions of magnetic materials with opposing polarities. Oscillating fluid flow was induced at a frequency of 100 Hz to enhance mixing performance. Split type design proved to have better mixing performance than the others.

Turbulence inside the fluidic chamber to improve mixing performance can also be produced by using pillars rotation as reported by Rahbar et al. [76]. Here, an individual or arrays of micromixer element in form of high aspect ratio of small pillar was fabricated using a micromolding process of nanomagnetic-composite polymers. The cilia, which are realized in PDMS (polydimethylsiloxane) doped with $(Nd_{0.7}Ce_{0.3})_{10.5}Fe_{83.9}B_{5.6}$ powder are then magnetized to produce permanent magnetic structures with bidirectional deflection capabilities, making them highly suitable as mixers controlled by electromagnetic fields. Similar to this concept, Zhou et al. [130] reported the development of magnetic pillars made of polymers composite with embedded Fe_3O_4 magnetic particles. Through an external magnetic field exposure, the magnetic pillar will react in rotation mode.

Furthermore, Pawinanto et al., [131] has developed polymer pillars on a movable flexible magnetic membrane with an attached permanent magnet (Figure 9, left). The movement of the pillars followed the deformation profile of the membrane (Figure 9, right). The concept of pillar rotations or membrane with pillar deformation in a mixer chamber has significantly increased the index of turbulence enabling higher mixing efficiency. These improvements thank to the advancement in the fabrication method of active micromixer that simplifies the mixer structure and its fabrication.

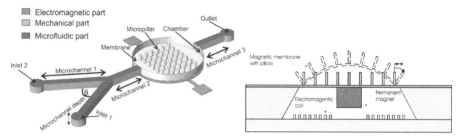

Figure 9. Schematic diagram of a pillar-based active microfluidic mixer device (**left**) and the micromixer structure showing the polymer pillars swing profile following the deformation of the actuator membrane (**right**).

5. Conclusions and Future Aspects of Polymers for EM Actuator

In this study, a comprehensive review on electromagnetically driven MEMS actuators with polymer-based movable structure is presented. The flexible characteristic of polymer is beneficial in attaining large and controlled structure deformation of the movable parts, such as thin membrane diaphragms, pillars or cantilevers. Significant discovery of polymer-based functional material has led to a wide range application of electromagnetic MEMS actuator. The flexible actuator structure with high magnetic property plays an important role in various biomedical instrumentations, such as lab on chip and drug delivery system.

These actuators can function as micropumps, microvalves and micromixers which execute the imperative roles of delivering biomedical samples. The wonderful combination between flexibility and magnetic properties of the magnetic polymer actuators can accurately control the microfluidic flow in a microchannel and determine its direction. In addition, the fluidic samples can be delivered precisely at a wide range of fluid flow rate, from 30 mL/min down to several nL/min. It will be a challenging effort to widen the flow rate range of an electromagnetic injection system which may require significant arrangement in pump and valve system design. This will eventually improve the reliability and quality of the electromagnetically driven microactuator system, specifically designed for drug delivery and artificial kidney.

The actuator structure also plays an important role as an active microfluidic mixer in the preparation process of the biomedical samples for drug delivery and lab on chip system. The polymer actuator can potentially reduce the mixing time and increase the mixing index. The increase of fluid sample turbulence inside the mixer chamber driven by external magnetic fields improves the mixing performance. Here, the innovation in design and fabrication technology for magnetic polymers introduces more compact mixer structure.

The presence of bulk permanent magnet attached onto the actuator has been identified as one of the main drawbacks for making an electromagnetically driven MEMS actuator to be large in size. Hence, it is crucial to make the actuator structure compact, as this will ultimately reduce the overall size of the system. A polymer membrane diaphragm with embedded magnetic nanoparticles can become a compact actuator with better mechanical properties. The developmental concept for magnetic actuator has evolved from the utilization of hard and fragile materials to more flexible polymeric materials with matrix magnet and now progresses towards embedded magnetic nanoparticles polymer composites. In the future, the polymer composites will eliminate the need of a conventional bulky permanent magnet in electromagnetic actuators.

To conclude, much progress has been made on magnetic actuator development and the future trend shows magnetic polymer composites as the new functional materials for flexible biomedical device technology. The magnetic polymer composite will be a fascinating material to be implemented in wearable and portable biomedical device applications that are currently and rapidly growing.

Author Contributions: J.Y. provided analysis and wrote article; M.M.S., R.E.P., W.A.F.W.A. and A.S. performed data collection, provided analysis and wrote article; B.M., I.H., A.A.H. and R.L. provided article review and editing; B.Y.M. supervised the article review. All authors have read and agreed to the published version of the manuscript.

Funding: This work is supported by AKU 254: HiCoE (Fasa II) 'MEMS for Biomedical Devices (Artificial Kidney) from the Ministry of Education Malaysia, the Ministry of Higher Education under LRGS Project Grant No. LRGS/2015/UKM-UKM/NANOMITE/04/01 and the Directorate of Research and Community Service, Ministry of Education and Culture, Republic of Indonesia.

Acknowledgments: The authors acknowledge grateful to Universitas Pendidikan Indonesia (UPI) and Universiti Kebangsaan Malaysia (UKM) for the joint research project in the field of MEMS sensors and actuators.

Conflicts of Interest: The authors declare no conflict of interest.

References

1. Fallahi, H.; Zhang, J.; Phan, H.P.; Nguyen, N.T. Flexible Microfluidics: Fundamentals, Recent Developments, and Applications. *Micromachines* **2019**, *10*, 830. [CrossRef]
2. Barth, C.; Knospe, C. Actuation of Flexible Membranes via Capillary Force: Single-Active-Surface Experiments. *Micromachines* **2018**, *9*, 545. [CrossRef]
3. Cordill, M.J.; Glushko, O.; Putz, B. Electro-Mechanical Testing of Conductive Materials Used in Flexible Electronics. *Front. Mater.* **2016**, *3*. [CrossRef]
4. Susheel, K.; Sarita, K.; Amit, K.; Yuvaraj, H.; Bandna, K.; Rajesh, K. Magnetic polymer nanocomposites for environmental and biomedical applications. *Colloid Polym. Sci.* **2014**, *292*, 2025–2052.
5. Imai, S.; Tsukioka, T. A magnetic MEMS actuator using a permanent magnet and magnetic fluid enclosed in a cavity sandwiched by polymer diaphragms. *Precis. Eng.* **2014**, *38*, 548–554. [CrossRef]
6. Safonovs, R. Design and Modelling of Electromagnetic Actuation in MEMS Switches, 2017. Master's Thesis, The University of Southern Denmark, Odense, Denmark, 2017.
7. Ashraf, M.W.; Tayyaba, S.; Afzulpurkar, N. Micro Electromechanical Systems (MEMS) Based Microfluidic Devices for Biomedical Applications. *Int. J. Mol. Sci.* **2011**, *12*, 3648–3704. [CrossRef]
8. Hilber, W. Stimulus-active polymer actuators for next-generation microfluidic devices. *Appl. Phys. A* **2016**, *122*, 751. [CrossRef]
9. Said, M.M.; Yunas, J.; Pawinanto, R.E.; Majlis, B.Y. PDMS based electromagnetic actuator membrane with embedded magnetic particles in polymer composite. *Sens. Actuators A* **2016**, *245*, 85–96. [CrossRef]
10. Hamid, N.A.; Ibrahim, M.; Radzi, S.A.; Chiew, W.Y.; Yunas, J.; Majlis, B.Y. A stack bonded thermo-pneumatic micro-pump utilizing polyimide based actuator membrane for biomedical applications. *Microsyst. Technol.* **2017**, *23*, 4037–4043. [CrossRef]
11. Liewellyn-Evans, H.; Griffiths, C.A.; Fahmy, A. Microgripper design and evaluation for automated l-wire assembly: A survey. *Microsyst. Technol.* **2020**. [CrossRef]
12. Singh, G.; Shahin, S.; Juliet, A.V. Efficient Low Frequency, Low Power Electromagnetically Actuated Acoustic Microspeaker for Hearing Aid Applications. *IOSR J. Electron. Commun. Eng. (IOSR-JECE)* **2015**, 15–20.
13. Brian, K.; Meng, E. Review of polymer MEMS micromachining. *J. Micromech. Microeng.* **2016**, *26*, 013001. [CrossRef]
14. Yamahata, C.; Chastellain, M.; Parashar, V.K.; Petri, A.; Hofmann, H.; Gijs, M.A.M. Plastic micropump with ferrofluidic actuation. *J. Microelectromech. Syst.* **2005**, *14*. [CrossRef]

15. Do, T.N.; Phan, H.; Nguyen, T.Q.; Visell, Y. Miniature Soft Electromagnetic Actuators for Robotic Applications. *Adv. Funct. Mater.* **2018**, *28*. [CrossRef]

16. Brauer, J.R. *Magnetic Actuators and Sensors*; IEEE Magnetic Society; John Willey & Sons Inc.: Hoboken, NJ, USA, 2006. [CrossRef]

17. Mi, S.; Pu, H.; Xia, S.; Sun, W. A Minimized Valveless Electromagnetic Micropump for Microfluidic Actuation on Organ Chips. *Sens. Actuators A Phys.* **2020**, *301*, 111704. [CrossRef]

18. Hamid, N.A.; Yunas, J.; Bahadorimehr, A.R.; Majlis, B.Y. Design Consideration of Membrane Structure for Thermal Actuated Micropump. *Adv. Mater. Res.* **2011**, *254*, 42–45. [CrossRef]

19. Zhou, Y. *Thesis Doctor of Philosophy in Mechanical Engineering*; The University of Illinois: Chicago, IL, USA, 2010.

20. Yaakub, T.N.T.; Yunas, J.; Latif, R.; Hamzah, A.A.; RazipWee, M.F.M.; Majlis, B.Y. Surface Modification of Electroosmotic Silicon Microchannel Using Thermal Dry Oxidation. *Micromachines* **2018**, *9*, 222. [CrossRef]

21. Bahadorimehr, A.R.; Jumril Yunas, J.; Majlis, B.Y. Low cost fabrication of microfluidic microchannels for Lab-On-a-Chip applications. In Proceedings of the IEEE Conference 2010 International Conference on Electronic Devices, Systems and Applications, Kuala Lumpur, Malaysia, 11–14 April 2010.

22. Feng, H.; Miao, X.; Yang, Z. Design, Simulation and Experimental Study of the Linear Magnetic Microactuator. *Micromachines* **2018**, *9*, 454. [CrossRef]

23. Venstra, W.J.; Sarro, P.M. Fabrication of crystalline membranes oriented in the (111) plane in a (100) silicon wafer. *Microelectron. Eng.* **2003**, *68*, 502–507. [CrossRef]

24. Tanaka, Y. Electric actuating valves incorporated into an all glass-based microchip exploiting the flexibility of ultra thin glass. *RSC Adv.* **2013**, *3*, 10213–10220. [CrossRef]

25. Korlyakov, A.V.; Mikhailova, O.N.; Serkov, A.V. Metallic coatings for MEMS structures. *IOP Conf. Ser. Mater. Sci. Eng.* **2018**, *387*, 012040. [CrossRef]

26. Chen, Y.C.; Kohl, P.A. Photosensitive sacrificial polymer with low residue. *Microelectron. Eng.* **2011**, *88*, 3087–3093. [CrossRef]

27. Abel, L.T.; Rodney, S.R.; Melody, A.S.; Matthew, R.G. An ultra-thin PDMS membrane as a bio/micro–nano interface: Fabrication and characterization. *Biomed. Microdevices* **2007**, *9*, 587–595.

28. Suter, M.; Ergeneman, O.; Schmid, S.; Camenzind, A.; Nelson, B.J.; Hierold, C. Supermagnetic Photosensitive Polymer Nanocomposite for Microactuators. In Proceedings of the TRANSDUCERS 2009—2009 International Solid-State Sensors, Actuators and Microsystems Conference, Denver, CO, USA, 21–25 June 2009.

29. Alvankarian, J.; Bahadorimehr, A.; Majlis, B.Y. A pillar-based microfilter for isolation of white blood cells on elastomeric substrate. *Biomicrofluidics* **2013**, *7*, 014102. [CrossRef] [PubMed]

30. Yue, F.G.; Zhao, Y. Microstructured Actuation of Liquid Crystal Polymer Networks. *Adv. Funct. Mater.* **2020**, *30*, 1901890. [CrossRef]

31. Zhang, R.Q.; Hong, S.L.; Wen, C.Y.; Pang, D.Y.; Zhang, Z.I. Rapid detection and subtyping of multiple influenza viruses on a microfluidic chip integrated with controllable micro-magnetic field. *Biosens. Bioelectron.* **2018**, *100*, 348–354. [CrossRef]

32. Masrie, M.; Yunas, J.; Majlis, B.Y.; Dehzangi, A. Vertically integrated optical transducer for bio-particle detection. *J. Eng. Sci. Technol.* **2017**, *12*, 1886–1899.

33. Bar-Cohen, Y.; Anderson, I.A. Electroactive polymer (EAP) actuators—Background review. *Mech. Soft Mater.* **2019**, *1*, 5. [CrossRef]

34. Gao, N.; Hou, G.; Liu, J.; Shen, J.; Gao, Y.; Zhang, L. Tailoring the mechanical properties of polymer nanocomposites via interfacial engineering. *Phys. Chem. Chem. Phys.* **2019**, *21*, 18714–18726.

35. Praveen, K.M.; Pious, C.V.; Thomas, S.; Grohens, Y. *Non-Thermal Plasma Technology for Polymeric Materials, Applications in Composites, Nanostructured Mateiasl and Biomedical Fields*; 2019; pp. 1–21. [CrossRef]

36. Grujic, A.; Talijan, N.; Stojanovic, D.; Stajić-Trosic, J.; Burzic, Z.; Balanovic, L.; Aleksic, R. Mechanical and magnetic properties of composite materials with polymer matrix. *J. Min. Metall. Sect. B Metall.* **2010**, *46*, 25–32. [CrossRef]

37. Rekosova, J.; Dosoudil, R.; Usakova, M.; Usak, E.; Hudec, I. Magnetopolymer Composites with Soft Magnetic Ferrite Filler. *IEEE Trans. Magn.* **2013**, *49*, 38–41. [CrossRef]

38. Madou, M.J. *Fundamentals of Microfabrication: The Science of Miniaturization*, 2nd ed.; CRC Press: Boca Raton, FL, USA, 2000.

39. Masrie, M.; Majlis, B.Y.; Yunas, J. Fabrication of multilayer-PDMS based microfluidic device for bio-particles concentration detection. *Bio-Med Mater. Eng.* **2014**, *24*, 1951–1958. [CrossRef]

40. Tsao, C.W. Polymer Microfluidics: Simple, Low-Cost Fabrication Process Bridging Academic Lab Research to Commercialized Production. *Micromachines* **2016**, *7*, 225. [CrossRef] [PubMed]

41. Suter, M.; Li, Y.; Sotiriou, G.A.; Teleki, A.; Pratsinis, S.E.; Hierold, C. low-cost fabrication of pmma and pmma based magnetic composite cantilevers. In Proceedings of the 16th International Solid-State Sensors, Actuators and Microsystems Conference 2011, Beijing, China, 5–9 June 2010. [CrossRef]

42. Kilani, M.I.; Galambos, P.C.; Haik, Y.S.; Chen, C.J. Design and analysis of a surface micromachined spiral-channel viscous pump. *J. Fluids Eng.* **2003**, *125*, 339–344. [CrossRef]

43. Sim, W.; Oh, J.; Choi, B. Fabrication, experiment of a microactuator using magnetic fluid for micropump application. *Microsyst. Technol.* **2006**, *12*, 1085–1091. [CrossRef]

44. Sylgard® 184 Silicone Elastomer Datasheet. Available online: www.dowcorning.com (accessed on 24 December 2019).

45. Hamid, N.A.; Yunas, J.; Majlis, B.Y.; Hamzah, A.A.; Bais, B. Microfabrication of Si3N4-polyimide membrane for thermo-pneumatic actuator. *Microelectron. Int.* **2015**, *32*, 18–24. [CrossRef]

46. Shearwood, C.; Harradine, M.A.; Birch, T.S.; Stevens, J.C. Applications of polyimide membrane to MEMS technology. *Microelectron. Eng.* **1996**, *30*, 547–550. [CrossRef]

47. DUPONT(TM)KAPTON®, Datasheet. Available online: https://www.dupont.com/content/dam/dupont/products-and-services/membranes-and-films/polyimde-films/documents/DEC-Kapton-summary-of-properties.pdf (accessed on 24 December 2019).

48. Kim, K.H.; Yoon, H.J.; Jeong, O.C.; Yang, S.S. Fabrication and test of a micro electromagnetic actuator. *Sens. Actuators A Phys.* **2005**, *117*, 8–16. [CrossRef]

49. Zhou, Y. Design and Microfabrication of an Elctromagentically Actuated Soft Polymer Micropump. Master's Thesis, Beijing University of Technology, Beijing, China, 2006.

50. Nguyen, N.T.; Truong, T.Q. A fully polymeric micropump with piezoelectric actuators. *Sens. Actuators B Chem.* **2004**, *97*, 137–143. [CrossRef]

51. Cho, J.; Anderson, M.; Richards, R.; Bahr, D.; Richards, C. Optimization of electromechanical coupling for a thin-film PZT membrane: II. Experiment. *J. Micromech. Microeng.* **2005**, *15*, 1804–1809. [CrossRef]

52. Sun, Y.; Piyabongkarn, D.; Sezen, A.; Nelson, B.J.; Rajamani, R. A high-aspect-ratio two-axis electrostatic microactuator with extended travel range. *Sens. Actuators A Phys.* **2002**, *102*, 49–60. [CrossRef]

53. Francais, O.; Dufour, I. Enhancement of elementary displaced volume with electrostatically actuated diaphragms: Application to electrostatic micropumps. *J. Micromech. Microeng.* **2000**, *10*, 282–286. [CrossRef]

54. Pu, C.; Park, S.; Chu, P.B.; Lee, S.S.; Tsai, M.; Peale, D.; Bonadeo, N.H.; Brener, I. Electrostatic actuation of three-dimensional MEMS mirrors using sidewall electrodes. *IEEE J. Sel. Top. Quantum Electron.* **2004**, *10*, 472–477. [CrossRef]

55. Conrad, H.; Schenk, H.; Kaiser, B.; Langa, S.; Gaudet, M.; Enz, M. A small-gap electrostatic micro-actuator for large deflections. *Nat. Commun.* **2015**, *6*, 10078. [CrossRef]

56. Ni, J.H.; Li, B.Z.; Yang, J.G. A MEMS-Based PDMS Micropump Utilizing Electromagnetic Actuation and Planar In-Contact Check Valves. *Adv. Mater. Res.* **2010**, *139–141*, 1574–1577. [CrossRef]

57. Xia, F.; Xu, T.; Tadigadapa, S.; Zhang, Q.M. Electroactive polymers for microactuators and microfluidic devices. In Proceedings of the 7th International Conference on Miniaturized Chemical and Biochemical Analysis Systems, Squaw Valley, CA, USA, 5–9 October 2003; pp. 195–198.

58. Cooney, C.G.; Towe, B.C. A thermopneumatic dispensing micropump. *Sens. Actuator A Phys.* **2004**, *116*, 519–524. [CrossRef]

59. Amirouche, F.; Zhou, Y.; Johnson, T. Current micropump technologies and their biomedical applications. *Microsyst. Technol.* **2009**, *145*, 647–666. [CrossRef]

60. Yamahata, C.; Lotto, C.; Al, E.; Gijs, M.A.M. A PMMA valveless micropump using electromagnetic actuation. *Microfluid. Nanofluidics* **2005**, *2*, 197–207. [CrossRef]

61. Büttgenbach, S. Electromagnetic Micromotors-Design, Fabrication and Applications. *Micromachines* **2014**, *5*, 929–942. [CrossRef]

62. Lee, S.A.; Lee, H.; Pinney, J.R.; Khialeeva, E.; Bergsneider, M.; Judy, J.W. Development of Microfabricated Magnetic Actuators for Removing Cellular Occlusion. *J. Micromech. Microeng.* **2011**, *21*, 054006. [CrossRef]

63. Zhou, Y.; Amirouche, F. An electromagnetically-actuated all-PDMS valveless. *Micromachines* **2011**, *11*, 345–355. [CrossRef]

64. Nouri, D.; Zabihi-Hesari, A.; Passandideh-Fard, M. Rapid mixing in micromixers using magnetic field. *Sens. Actuators A Phys.* **2017**, *255*, 79–86. [CrossRef]

65. Liu, F.; Zhang, J.; Alici, G.; Yan, S.; Mutlu, R.; Li, W.; Yan, T. An inverted micro-mixer based on a magnetically-actuated cilium made of Fe doped PDMS. *Smart Mater. Struct.* **2016**, *25*, 95049. [CrossRef]

66. Banis, G.; Tyrovolas, K.; Angelopoulos, S.; Ferraro, A.; Hristoforou, E. Pushing of Magnetic Microdroplet Using Electromagnetic Actuation System. *Nanomaterials* **2020**, *10*, 371. [CrossRef] [PubMed]

67. Rinklin, P.; Krause, H.J.; Wolfrum, B. On-chip electromagnetic tweezers—3-dimensional particle actuation using microwire crossbar arrays. *Lab Chip* **2016**, *24*, 4749–4758. [CrossRef]

68. Chen, L.; Offenhäusser, A.; Rause, H.J. Magnetic tweezers with high permeability electromagnets for fast actuation of magnetic beads. *Rev. Sci. Instrum.* **2015**, *86*, 044701. [CrossRef]

69. Choi, J.W.; Hahn, C.A.; Bhansali, H.; Henderson, T. A new magnetic bead-based, filterless bio-separator with planar electromagnet surfaces for integrated bio-detection systems. *Sens. Actuators B Chem.* **2000**, *6*, 34–39. [CrossRef]

70. Thielicke, E.; Obermeier, E. Microactuators and Their Technologies. *Mechatronics* **2000**, *10*, 431–455. [CrossRef]

71. Pawinanto, R.E.; Yunas, J.; Said, M.M.; Majlis, B.Y.; Hamzah, A.A. Design Consideration of Planar Embedded Micro-Coils for Electromagnetic Actuator of Fluids Injection System. *Middle-East J. Sci. Res.* **2014**, *19*, 538–543. [CrossRef]

72. Pawinanto, R.E.; Yunas, J.; Majlis, B.Y.; Hamzah, A.A. Design and Fabrication of Compact MEMS Electromagnetic Micro-Actuator with Planar Micro-Coil Based on PCB. *Telkomnika* **2016**, *14*, 856–866. [CrossRef]

73. Engel, A.; Friedrichs, R. On the electromagnetic force on a polarizable body. *Am. J. Phys.* **2002**, *70*, 428432. [CrossRef]

74. Sugandi, G.; Yunas, J.; Hamzah, A.A.; Noor, M.M.; Wiranto, G.; Majlis, B.Y. Design, Fabrication and Characterization of Electrodynamically Actuated MEMS-Speaker. *ASM Sci. J.* **2019**, *12*, 125–130.

75. Abhari, F.; Jaafar, H.; Yunus, N.A. A Comprehensive Study of Micropump Technologies. *Int. J. Electrochem. Sci.* **2012**, *7*, 9765–9780.

76. Rahbar, M.; Shannon, L.; Gray, B.L. Microfluidic active mixers employing ultra-high aspect-ratio rare-earth magnetic nano-composite polymer artificial cilia. *J. Micromech. Microeng.* **2014**, *24*, 025003. [CrossRef]

77. Varadan, V.K. *Microelectromechanical Systems (MEMS)*; Wiley: Hoboken, NJ, USA, 2003; pp. 1–49.

78. Dai, X.; Xie, H. A simple and residual-layer-free solute- solvent separation soft lithography method. *J. Micromech. Microeng.* **2015**, *25*, 10–18. [CrossRef]

79. Pramanick, B.; Dey, P.K.; Das, S.; Bhattacharya, T.K. Design and Development of a PDMS Membrane based SU-8 Micropump for Drug Delivery System. *J. ISSS* **2013**, *2*, 1–9.

80. Ni, J.; Wang, B.; Chang, S.; Lin, Q. An integrated planar magnetic micropump. *Microelectron. Eng.* **2013**, *117*, 35–40. [CrossRef]

81. Munas, F.R.; Amarasinghe, Y.W.; Dao, D. Review on MEMS based Micropumps for Biomedical Applications. *Int. J. Innov. Res. Sci. Eng. Technol.* **2015**, *4*, 5602–5615.

82. Zhang, R.; You, F.; Lv, Z.; He, Z.; Wang, H.; Huang, L. Development and Characterization a Single-Active-Chamber Piezoelectric Membrane Pump with Multiple Passive Check Valves. *Sensors* **2016**, *16*, 2108. [CrossRef]

83. Wu, C.H.; Chen, C.W.; Kuo, L.S.; Chen, P.H. A Novel Approach to Measure the Hydraulic Capacitance of a Microfluidic Membrane Pump. *Adv. Mater. Sci. Eng.* **2014**. [CrossRef]

84. Qin, D.; Xia, Y.; Whitesides, G.M. Soft lithography for micro- and nanoscale patterning. *Nat. Protoc.* **2010**, *5*, 491–501. [CrossRef] [PubMed]

85. Au, A.K.; Lai, H.; Ben, R.; Folch, A. Microvalves and Micropumps for BioMEMS. *Micromachines* **2011**, *2*, 179–220. [CrossRef]

86. Ghanbari, A.; Nock, V.; Johari, S.; Blaikie, R.; Chen, X.; Wang, W. A micropillar-based on-chip system for continuous force measurement of C. elegans. *J. Micromech. Microeng.* **2012**, *22*, 095009. [CrossRef]

87. Yunas, J.; Pawinnto, R.E.; Indah, N.; Alva, S.; Sebayang, D. The Electrical and Mechanical Characterization of Silicon Based Electromagnetic Microactuator for Fluid Injection System. *J. Eng. Sci. Technol.* **2018**, *13*, 2606–2615.

88. Said, M.M.; Yunas, J.; Bais, B.; Hamzah, A.A.; Majlis, B.Y. Electromagnetic micrpump with a matrix-patterned magnetic polymer composite actuator membrane. *Micromachines* **2018**, *9*, 1–10.

89. Xue, W.; Cui, T. Polymer Magnetic Microactuators Fabricated with Hot Embossing and Lyer-by-Layer Nano Self-Assembly. *J. Nanosci. Nanotechnol.* **2007**, *7*, 2647–2653. [CrossRef]

90. Low, Z.X.; Chua, Y.T.; Ray, B.M.; Mattia, D.; Metcalfe, I.S.; Patterson, D.A. Perspective on 3D printing of separation membranes and comparison to related unconventional fabrication techniques. *J. Membr. Sci.* **2017**, *523*, 596–613. [CrossRef]

91. Amin, R.; Joshi, A.; Tasoglu, S. Commercialization of 3D-printed microfluidics devices. *J. 3D Print. Med.* **2017**, *1*, 85–89. [CrossRef]

92. Lee, Y.S.; Bhattacharjee, N.; Folch, A. 3D-printed Quake-style microvalves and micropumps. *Lab Chip* **2018**, *8*, 1207–1214. [CrossRef]

93. Zhou, Z.; He, G.; Zhang, K.; Zhao, Y.; Sun, D. 3D-Printed membrane microvalves and microdecoder. *Microsyst. Technol.* **2019**, *25*, 4019–4025. [CrossRef]

94. Kim, S.; Lee, J.; Choi, B. 3D Printed Fluidic Valves for Remote Operation via External Magnetic Field. *Int. J. Precis. Eng. Manuf.* **2016**, *17*, 937–942. [CrossRef]

95. Au, A.K.; Bhattacharjee, N.; Horowitz, L.S.; Chang, T.C.; Folch, A. 3D-Printed Microfluidic Automation. *Lab Chip* **2015**, *15*, 1934–1940. [CrossRef] [PubMed]

96. Thevenota, J.; Oliveira, H.; Sandre, O.; Lecommandoux, S. Magnetic responsive polymer composite materials. *Chem. Soc. Rev.* **2013**, *42*, 7099–7116. [CrossRef] [PubMed]

97. Yeh, Y.H.; Cho, K.H.; Chen, L.J. Effect of the softness of polydimethylsiloxine on the hydrophobicity of pillar-like patterned surfaces. *Soft Matter.* **2012**, *8*, 1079. [CrossRef]

98. Tahmasebipour, M.; Paknahad, A.A. Unidirectional and bidirectional valveless electromagnetic micropump with PDMS-Fe3O4 nanocomposite magnetic membrane. *J. Micromech. Microeng.* **2019**, *29*, 075014. [CrossRef]

99. Zhang, Q.Y.; Zhang, H.P.; Xie, G.; Zhang, J.P. Effect of surface treatment of magnetic particles on the preparation of magnetic polymer microspheres by miniemulsion polymerization. *J. Magn. Magn. Mater.* **2007**, *311*, 140–144. [CrossRef]

100. Lu, A.H.; Salabas, E.L.; Schuth, F. Magnetic nanoparticles: Synthesis, protection, functionalization, and application. *Angew. Chem.-Int. Ed.* **2007**, *46*, 1222–1244. [CrossRef]

101. Suter, M.; Ergeneman, O.; Zürcher, J.; Moitzi, C.; Pané, S.; Rudin, T.; Pratsinis, S.E.; Nelson, B.J.; Hierold, C. A photopatternable superparamagnetic nanocomposite: Material characterization and fabrication of microstructures. *Sens. Actuators B Chem.* **2011**, *156*, 433–443. [CrossRef]

102. Fahrni, F. Magnetic Polymer Actuators for Microfluidics. Ph.D. Thesis, Technische Universiteit Eindhoven, Eindhoven, The Netherland, 2009. [CrossRef]

103. Liu, C. Recent developments in polymer MEMS. *Adv. Mater.* **2007**, *19*, 3783–3790. [CrossRef]

104. Rahbar, M. Design, Fabrication and Testing of Magnetic Composite Polymer Actuators Integrated With Microfluidic Devices and Systems. Ph.D. Thesis, Simon Fraser University, Burnaby, BC, Canada, 2016.

105. Schneider, F.; Draheim, J.; Müller, C.; Wallrabe, U. Optimization of an adaptive PDMS-membrane lens with an integrated actuator. *Sens. Actuators A* **2009**, *154*, 316–321. [CrossRef]

106. Dai, Q.; Berman, D.; Virwani, K.; Frommer, J.; Jubert, P.O.; Lam, M.; Topuria, T.; Imaino, W.; Nelson, A. Self-assembled ferrimagnet–polymer composites for magnetic recording media. *Nano Lett.* **2010**, *10*, 3216–3221. [CrossRef] [PubMed]

107. Cho, H.J.; Ahn, C.H. A bidirectional magnetic microactuator using electroplated permanent magnet arrays a bidirectional magnetic microactuator using electroplated permanent magnet arrays. *J. Microelectromicromechanical Syst.* **2002**, *11*, 78–84. [CrossRef]

108. Su, Y.; Chen, W. Investigation on electromagnetic microactuator and its application in Micro-Electro-Mechanical System (MEMS). In Proceedings of the 2007 IEEE International Conference on Mechatronics and Automation (ICMA), Harbin, China, 5–8 August 2007; pp. 3250–3254.

109. Samaniego, L.F.P. Design Magnetically Actuated Surgical Devices Using Magnetic Micro-Particles Embedded in a Polymeric Matrix. Master's Thesis, University of Groningen, Groningen, The Netherlands, 2018.

110. Dogangil, G.; Davies, B.L.; Baena, F.R. A review of medical robotics for minimally invasive soft tissue surgery. *Proc. Inst. Mech. Eng. Part H J. Eng. Med.* **2010**, *224*, 653–679. [CrossRef] [PubMed]

111. Zhang, S.; Wang, Y.; Lavrijsen, R.; Onck, P.R.; den Toonder, J.M.J. Versatile microfluidic flow generated by moulded magnetic artificial cilia. *Sens. Actuators B Chem.* **2018**, *263*, 614–624. [CrossRef]

112. Johari, J.; Yunas, J.; Hamzah, A.A.; Majlis, B.Y. Piezoelectric Micropump with Nanoliter Per Minute Flow for Drug Delivery Systems. *Sains Malays.* **2011**, *40*, 275–281.

113. Wang, Y.H.; Tsai, Y.W.; Tsai, C.H.; Lee, C.Y.; Fu, L.M. Design and analysis of impedance pumps utilizing electromagnetic actuation. *Sensors* **2010**, *10*, 4040–4052. [CrossRef]

114. Pan, T.; McDonald, S.J.; Kai, E.M.; Ziaie, B. A magnetically driven PDMS micropump with ball check-valves. *J. Micromech. Microeng.* **2005**, *15*, 1021–1026. [CrossRef]

115. Yin, H.; Huang, Y.; Fang, W.; Hsieh, J. A novel electromagnetic elastomer membrane actuator with a semi-embedded coil. *Sens. Actuators A Phys.* **2007**, *139*, 194–202. [CrossRef]

116. Nagel, J.J.; Mikhail, G.; Noh, H.M.; Koo, J. Magnetically actuated micropumps using an Fe-pdms composite membrane. *SPIE Smart Struct. Mater.* **2006**, *96*, 234–243.

117. Shen, C.; Liu, H. Innovative composite pdms micropump with electromagnetic drive. *Sens. Mater. J.* **2010**, *22*, 85–100.

118. Said, M.M.; Yunas, J.; Bais, B.; Hamzah, A.A.; Majlis, B.Y. Hybrid polymer composite membrane for an electromagnetic (EM) valveless micropump. *J. Micromech. Microeng.* **2017**, *27*, 075027. [CrossRef]

119. Sadler, D.J.; Oh, K.W.; Ahn, C.H.; Bhansali, S.; Henderson, H.T. A new magnetically actuated microvalve for liquid and gas control applications. In Proceedings of the 10th International Conference on Solid-State Sensors and Actuators (Transducers '99), Sendai, Japan, 7–10 June 1999; pp. 1812–1815.

120. Attila Gaspar, A.; Piyasena, M.E.; Daróczi, L.; Gomez, F.A. Magnetically controlled valve for flow manipulation in polymer microfluidic devices. *Microfluid. Nanofluid.* **2008**, *4*, 525–531. [CrossRef]

121. Bute, M.G.; Sheikh, A.; Mathe, V.L.; Bodas, D.; Karekar, R.N.; Gosavi, S.W. Magnetically controlled flexible valve for flow manipulation in polymer microfluidic devices. In Proceedings of the 1st International Symposium on Physics and Technology of Sensors (ISPTS-1), Pune, India, 7–10 March 2012.

122. Nakahara, T.; Suzuki, J.; Hosokawa, Y.; Suzuki, T. Fabrication of Magnetically Driven Microvalve Arrays Using a Photosensitive Composite. *Magnetochemistry* **2018**, *4*, 7. [CrossRef]

123. Dario, P.; Croce, N.; Carrozza, M.C.; Varallo, G. A fluid handling system for a chemical microanalyzer. *J. Micromech. Microeng.* **1996**, *6*, 95–98. [CrossRef]

124. Bohm, S.; Olthuis, W.; Bergveld, P. A plastic micropump constructed with conventional techniques and materials. *Sens. Actuators A* **1999**, *77*, 223–228. [CrossRef]

125. Khoo, M.; Liu, C. A novel micromachined magnetic membrane microfluid pump. In Proceedings of the 22nd Annual International Conference of the IEEE Engineering in Medicine and Biology Society, Chicago, IL, USA, 23–28 July 2000; pp. 2394–2397.

126. Shen, M.; Yamahata, C.; Gijs, M.A. A high-performance compact electromagnetic actuator for a PMMA ball-valve micropump. *J. Micromech. Microeng.* **2008**, *18*, 025031. [CrossRef]

127. Dich, N.Q.; Dinh, T.X.; Pham, P.H.; Dau, V.T. Study of valveless electromagnetic micropump by volume-of-fluid and open FOAM. *Jpn. J. Appl. Phys.* **2015**, *54*, 057201. [CrossRef]

128. Cai, G.; Xue, L.; Zhang, H.; Lin, J. A Review on Micromixers. *Micromachines* **2017**, *8*, 274. [CrossRef]

129. Tang, S.Q.; Li, K.H.H.; Yeo, Z.T.; Chan, W.X.; Tan, S.H.; Yoon, Y.J.; Ng, S.H. Study of concentric, eccentric and split type magnetic membrane micro-mixers. *Sens. Bio-Sens. Res.* **2018**, *19*, 14–23. [CrossRef]

130. Zhou, B.; Xu, W.; Syed, A.A.; Chau, Y.; Chen, L.; Chew, B.; Yassine, O.; Wu, X.; Gao, Y.; Zhang, J.; et al. Design and fabrication of magnetically functionalized flexible micropillar arrays for rapid and controllable microfluidic mixing. *Lab Chip* **2015**, *15*, 2125–2132. [CrossRef]

131. Pawinanto, R.E.; Yunas, J.; Hashim, A.M. Design optimization of active microfluidic mixer incorporating micropillar on flexible membrane. *Microsyst. Technol.* **2019**, *25*, 1203–1209. [CrossRef]

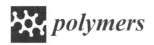

Article

Dynamic Mussel-Inspired Chitin Nanocomposite Hydrogels for Wearable Strain Sensors

Pejman Heidarian [1], Abbas Z. Kouzani [1,*], Akif Kaynak [1], Ali Zolfagharian [1] and Hossein Yousefi [2]

[1]   School of Engineering, Deakin University, Geelong, Victoria 3216, Australia;
      pheidarian@deakin.edu.au (P.H.); akif.kaynak@deakin.edu.au (A.K.); a.zolfagharian@deakin.edu.au (A.Z.)
[2]   Department of Wood Engineering and Technology, Gorgan University of Agricultural Sciences and Natural
      Resources, Gorgan 4913815739, Iran; hyousefi.ir@gmail.com
*   Correspondence: kouzani@deakin.edu.au

Received: 9 June 2020; Accepted: 23 June 2020; Published: 24 June 2020

Abstract: It is an ongoing challenge to fabricate an electroconductive and tough hydrogel with autonomous self-healing and self-recovery (SELF) for wearable strain sensors. Current electroconductive hydrogels often show a trade-off between static crosslinks for mechanical strength and dynamic crosslinks for SELF properties. In this work, a facile procedure was developed to synthesize a dynamic electroconductive hydrogel with excellent SELF and mechanical properties from starch/polyacrylic acid (St/PAA) by simply loading ferric ions (Fe^{3+}) and tannic acid-coated chitin nanofibers (TA-ChNFs) into the hydrogel network. Based on our findings, the highest toughness was observed for the 1 wt.% TA-ChNF-reinforced hydrogel (1.43 MJ/m^3), which is 10.5-fold higher than the unreinforced counterpart. Moreover, the 1 wt.% TA-ChNF-reinforced hydrogel showed the highest resistance against crack propagation and a 96.5% healing efficiency after 40 min. Therefore, it was chosen as the optimized hydrogel to pursue the remaining experiments. Due to its unique SELF performance, network stability, superior mechanical, and self-adhesiveness properties, this hydrogel demonstrates potential for applications in self-wearable strain sensors.

Keywords: dynamic hydrogels; tannic acid; chitin nanofibers; starch; self-healing; self-recovery

1. Introduction

Hydrogels are hydrophilic polymers cross-linked mostly by static covalent bonds in a three-dimensional (3D) structure [1–4]. They can maintain a large amount of water without losing their structures, and are suitable for many applications, including sensors [5], scaffolds [6], wound healing substrates [7], and actuators [8]. However, due to the presence of static bonds, they are usually prone to permanent failure while under load before any noticeable cracks appear, thus losing their functionality [1,2,9]. The insertion of dynamic non-covalent crosslinks within their networks can be considered as one feasible way to fabricate hydrogels with the ability to restore their structures and functionalities from damage, thus improving their safety, reliability, and durability. Furthermore, the reversibility of dynamic crosslinks in such hydrogels also imparts another interesting feature to their networks: autonomous self-healing and self-recovery (SELF) properties [1,3,10,11]. Therefore, dynamic hydrogels are good candidates for preparing soft flexible electronics, biomedicine, and wearable strain sensors [3,4,12,13]. However, the insertion of dynamic crosslinks may reduce the toughness of electroconductive hydrogels, the main prerequisite for fabricating hydrogel-based strain sensors that are subjected to repeated deformations [2,9]. Therefore, it is an ongoing challenge to fabricate an electroconductive hydrogel with toughness and autonomous SELF behaviors that are sufficiently suitable for wearable strain sensors because of the compromise between static crosslinks for mechanical strength and dynamic crosslinks for SELF properties [1,9,14,15].

Polymers **2020**, *12*, 1416

The current solution to fabricate a tough SELF hydrogel relies on embedding dynamically modified nanofillers, e.g., nanoclay, graphene oxide, carbon nanotubes, and nanocellulose, within the network of hydrogels containing covalently cross-linked bonds [16–18]. By doing so, nanofillers with dynamic motifs increase the binding affinity at the interface of polymer chains and enhance the energy dissipation within the structure of hydrogels. However, due to employing irreversible covalent crosslinks, a full restoration of damaged hydrogels is not feasible. As an example, Shao et al. [18] employed tannic acid-coated cellulose nanocrystals (TA-CNCs) into a covalently cross-linked polyacrylic acid-aluminum-ion (PAA-Al^{3+}) hydrogel to impart both SELF and mechanical strength to PAA hydrogel. Herein, TA, as a non-toxic, biocompatible plant-based polyphenol, provided strong metal-phenolic networks with Al^{3+} ions, thus imparting both SELF and mechanical strength to the PAA hydrogel; however, the presence of static bonds in the structure of the hydrogel restricted the fabrication of a fully reversible hydrogel [18].

In this work, TA-coated chitin nanofibers (TA-ChNFs) were employed as dynamic motifs for bestowing SELF and mechanical strength to a starch-based hydrogel without using any static bonds. ChNFs are highly crystallized fibrous structures that are mainly found in the exoskeleton of arthropods, e.g., crabs, shrimp, and insects [19,20]. They are formed linearly by the synthesis of glucosamine monomers connected by β-(1-4)-N-acetyl glucosamine linkages with an approximate diameter within a range of 2–20 nm [21]. ChNFs, similar to CNCs, have a good modifiability and provide excellent mechanical strength to hydrogels. Therefore, it is believed that TA-ChNFs can provide a high level of dynamic crosslinks between their adjacent nanofibers and polymer networks, thereby imparting both SELF and mechanical properties at the same time to a hydrogel network.

To fabricate hydrogels, polysaccharides are usually the most commonly used hydrophilic polymers because they are cheap, cytocompatible, biocompatible, and biodegradable, and among them, starch (St) is the most inexpensive and readily available polysaccharide [1]. In contrast with cellulose and chitin, which contain linear chains, St has highly-branched portions (amylopectin) with α(1,4)-anhydroglucose chains interlinked with α-(1 → 6)-glycosidic bonds, in association with some linear portions (amylose) with α(1,4)-linked anhydroglucose units [22]. As such, the presence of highly-branched chains in St always results in brittle and moisture-sensitive products with a poor mechanical strength. Therefore, it is almost impossible to use St in load-bearing applications, e.g., wearable strain sensors, due to its limited flexibility and stretchability. Herein, an electroconductive, tough hydrogel based on St with unique SELF properties was fabricated, suitable for wearable strain sensors.

Using such a hydrogel for wearable strain sensors also requires an external glue for fixing sensors onto substrates to prevent the interfacial delamination between the contacted substrates and sensors. This complicates the fabrication of hydrogels [23]. Therefore, there is a substantial need for developing self-adhesive soft wearable strain sensors. Polydopamine (PDA) is usually the main candidate for this additional feature, but its high cost and dark color may not always be useful for practical applications. In this regard, TA appears to be a better candidate than PDA because of its low-cost, non-toxic, nonirritant to human skin, and biocompatible plant-based nature [24,25]. Herein, by employing TA, a mussel-inspired self-adhesive performance was also added to the hydrogel. This resolves the need for using external glue to attach the hydrogel onto the contact substrate, thus avoiding interfacial delamination under a repeated deformation state and improving the stability of the signal detection.

2. Materials and Methods

2.1. Chemicals and Materials

Mechanically isolated chitin nanofibers (ChNFs) were supplied by Nano Novin Polymer Co. (Sari, Iran). Tannic acid (TA), acrylic acid (AA), starch (St), ammonium persulfate (APS), Tris buffer solution, and ferric chloride hexahydrate (FeCl$_3$·6H$_2$O) were purchased from Sigma-Aldrich (Castle Hill, Australia).

Polymers **2020**, *12*, 1416

2.2. TA-ChNFs Preparation

The suspension of TA-ChNFs was prepared by a procedure proposed by Shao et al. [18]. In brief, the pH was first adjusted to 8.0 by dropwise adding Tris buffer solution into 150 cc of ChNFs suspension (~1 wt.%). After that, 0.51 g of TA was loaded into the ChNFs suspension. The suspension was then magnetically stirred for 6 h at room temperature to coat the surface of ChNFs with TA. Next, the suspension was purified by a repeated centrifugation, followed by redispersing in distilled water. By doing so, the color of the suspension was changed from white to brown. The TA-ChNFs suspension was then sealed and cryopreserved at 4 °C.

2.3. Hydrogel Preparation

The hydrogels were fabricated based on the method developed by Hussain et al. [26] for fabricating glycogen-PAA hydrogels. In brief, 2 g of St was dissolved in 24 cc distilled water at 80 °C for 2 h, followed by cooling it down to room temperature. Next, 50 mg APS was added into the dissolved St to activate the functional groups of St and stirred for 10 min. After that, 4 g of AA was added to the mixture and stirred for another 10 min. Finally, 0.1 M ferric (Fe^{3+}) ions were loaded to the mixture, and the mixture was stirred again for 10 min at room temperature. TA-ChNF-loaded hydrogels were also fabricated at different concentrations (0.1, 0.5, 1, 1.5, and 2 wt.%) by loading TA-ChNFs into the dissolved St and sonicating for further 10 min in an ice-water bath to form a uniform mixture, followed by repeating the mentioned procedure for loading APS, AA, and Fe^{3+} ions. As seen in Scheme 1, after inclusion of TA-ChNF into St (Scheme 1b), AA monomers were polymerized using APS to impart a higher molecular polarity to starch/TA-ChNFs (Scheme 1c–e). Afterward, 0.1 M solution of Fe^{3+} ions was loaded into the St/PAA/TA-ChNFs mixture while mixing for almost 10 s at room temperature (Scheme 1f). Finally, superfluous cations from the Fe^{3+}-loaded hydrogels were removed by soaking them in deionized water for 24 h. All samples were coded according to St/PAA/TA-ChNF (x%), which is the weight ratio of the TA-ChNF against AA/St (constant at 6 g). As an example, 1.0 wt.% TA-ChNF hydrogel refers to the hydrogel containing 60 mg of TA-ChNF and is coded as St/PAA/TA-ChNF (1%). Table 1 shows the detailed composition of the hydrogels, and Scheme 1g shows the possible coordination modes after formation of the hydrogel.

Table 1. Compositions of the hydrogels.

Code	TA-ChNFs (wt.%)	TA-ChNFs (mg)	AA (g)	St (g)	APS (mg)	Water (cc)
St/PAA	0	0	4	2	50	24
St/PAA/TA-ChNF (0.1%)	0.1	6	4	2	50	24
St/PAA/TA-ChNF (0.5%)	0.5	30	4	2	50	24
St/PAA/TA-ChNF (1%)	1	60	4	2	50	24
St/PAA/TA-ChNF (1.5%)	1.5	90	4	2	50	24
St/PAA/TA-ChNF (2%)	2	120	4	2	50	24

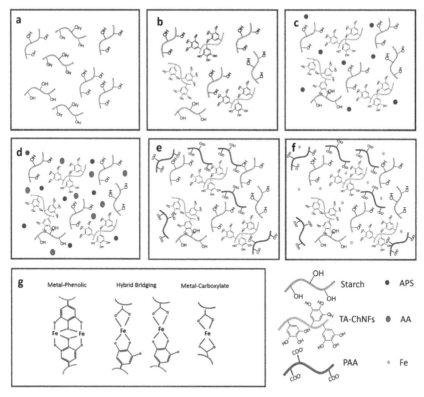

Scheme 1. (**a–f**) Different steps for forming St/PAA/TA-ChNFs hydrogels; (**g**) different possible modes of coordination crosslinks.

2.4. Hydrogel Characterization

The morphologies of ChNFs and TA-ChNFs were observed using transmission electron microscopy, TEM (JEOL, Tokyo, Japan, 2100). Each sample was diluted to about 0.001 wt.% with ethanol and sonicated well to separate the individual nanofibers and avoid aggregation during the analysis. Then, a quantity of 5 µL of the ChNFs was cast on perforated carbon-coated grids, and the excess ethanol was absorbed by a filter paper. An image analyzer program (version 1.52t, ImageJ, Bethesda, MD, USA) was used to measure the diameter of the nanofibers from the TEM images.

Next, the mechanical tests were performed using a universal mechanical tester equipped with a 200 N load cell at room temperature (Instron, Norwood, MA, USA) to find the hydrogel with the highest toughness at an optimum TA-ChNFs concentration. To do so, rectangular-shaped specimens, with dimensions of 10 mm in width, 6 mm in depth, and 35 mm in length, were prepared to test the mechanical properties of the TA-ChNF-loaded hydrogels at different concentrations (0.1, 0.5, 1, 1.5, and 2 wt.%). The constant stretching rate and initial distance between two clamps were respectively 60–160 mm/min and 15 mm. Prior to the test, the surface of hydrogels was coated with a layer of silicone oil to minimize water evaporation. The fracture strains of hydrogels were determined from the elongation at the break values. The toughness, that is the ability of the hydrogel to absorb energy prior to fracture, was calculated from the area under the stress–strain curves. All tensile tests were repeated five times. Rheological measurements were also performed to support the mechanical characterization using a rotational rheometer (TA Instruments, New castle, Delaware, USA) equipped with a parallel plate geometry (40 mm in diameter and a gap of 49 µm) at different strains and a frequency of 1.0 Hz.

After determining the hydrogel with the highest toughness, further experiments were performed on the samples with the optimized TA-ChNFs concentration. Scanning electron microscopy, SEM (Zeiss, Supra, Oberkochen, Germany), was employed to investigate the morphology of the freeze-dried hydrogel before and after loading TA-ChNFs. To do so, the optimized hydrogel was cut to expose the inner structure and coated with gold in a vacuum coater to avoid charging and then observed under SEM operating at 8.5 and 17.5 kV. The spectra of ChNFs, TA-ChNFs, and the hydrogels at the optimized TA-ChNFs concentration were obtained using FTIR (Bruker Vertex 70, Billerica, MA, USA) between 600–4000 cm^{-1}. Prior to the test, the samples were dried in a vacuum oven for 24 h.

The healing efficiency of hydrogels was also calculated according to the strength ratio between the healed hydrogel and the original one at different TA-ChNFs concentrations (0.1, 0.5, 1, 1.5, and 2 wt.%). Furthermore, a visual inspection was conducted to evaluate the self-healing properties of the hydrogel at an optimum TA-ChNFs concentration by cutting the hydrogel into pieces, followed by immediately recombining the pieces into the original shape through self-adhesion.

The self-adhesiveness properties of the hydrogel were measured at an optimum TA-ChNFs concentration on different substrates (plastic, glass, metal, leather, and rubber with $25 \times 100 \times 1$ mm^3 dimensions) using the same universal mechanical tester under ambient conditions. For this purpose, all solid specimens were washed with water and ethanol and dried to remove any dirt from their surfaces. Next, the hydrogels (with $20 \times 20 \times 1$ mm^3 dimensions) were attached to the substrates and were pulled at a crosshead speed of 10 mm/min until separation. Each sample for measuring the self-adhesive test was repeated five times [27].

3. Results and Discussion

3.1. Tannic Acid Coated-Chitin Nanofiber-Assisted Hydrogels

According to the TEM image of ChNFs (Figure 1a), the average diameter of ChNFs is in the nanosize range (48 ± 12 nm), while their lengths are in the micrometer scale. Therefore, ChNFs have a high aspect ratio containing a lot of hydroxyl groups able to interact with the pyrogallol/catechol groups of TA. Based on the TEM image (Figure 1b), the average diameter of TA-ChNFs is slightly thicker than ChNFs (51 ± 11 nm), which may be due to the deposition of TA onto ChNFs. The successful deposition of TA on ChNFs was confirmed using the FTIR test. As shown in Figure 1c, unlike the pristine ChNFs (Figure 1ci), there is a detectable peak at 813 cm^{-1} in the TA-ChNFs spectrum (Figure 1cii) due to the introduction of C=C in benzene rings. There are also detectable peaks at 1531 and 1612 cm^{-1} due to stretching vibrations of C−C aromatic groups in the spectrum of TA-ChNFs. These results are in agreement with the results obtained by Shao et al. [18] when depositing TA on the surfaces of cellulose nanocrystals.

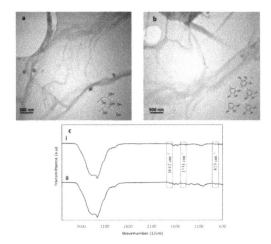

Figure 1. TEM image of (**a**) ChNFs and (**b**) TA-ChNFs; and FTIR results of (**ci**) ChNFs and (**cii**) TA-ChNFs.

Inspired by the adhesiveness of blue mussels, a tough hydrogel with SELF properties was fabricated via the incorporation of TA-ChNFs into St/PAA, followed by loading Fe^{3+} ions. The nearly instant gelation of 0.5 wt.% TA-ChNF-reinforced St/PAA is shown in Video S1 from Supplementary Materials. The instant gelation was not observed for the mixture containing 0.1 wt.% TA-ChNFs (Video S2). The instant gelation can be attributed to the sufficient contribution of pyrogallol/catechol groups of TA performing intermolecular interactions with Fe^{3+} ions, thus forming coordination crosslinks. In fact, after loading Fe^{3+} ions into the TA-ChNFs-assisted mixture, three possible coordination modes are likely to happen, consisting of hybrid bridging, metal-carboxylate coordination, and metal-phenolic coordination (Scheme 1g). All of these modes, as well as the presence of hydrogen bonding, result in a reversible dynamic network that can instantly form a 3D gel.

3.2. Mechanical and Rheological Properties

Since toughness is one of the most important prerequisites of a wearable strain sensor, mechanical measurements were first performed on all hydrogels at different TA-ChNFs concentrations (0.1, 0.5, 1, 1.5, and 2 wt.%) to indicate the hydrogel with the highest toughness. As mentioned, St is inherently a brittle polymer, so the graft polymerization of vinyl monomers, e.g., acrylic acid and acrylamide, can be a facile way to modify the mechanical strength of St. Moreover, the incorporation of nanofillers within the St network can improve the mechanical strength of St [28]. Therefore, it is anticipated that the incorporation of TA-ChNFs and the graft polymerization of AA without using any static cross-linkers can considerably increase the toughness of St-based hydrogels while imparting excellent SELF properties to them. These two properties (toughness and SELF properties) usually oppose each other [9]. Figure 2a displays the stress–strain curves of the hydrogels incorporated with different concentrations of TA-ChNFs (0.1, 0.5, 1, 1.5, and 2 wt.%). As can be seen, the incorporation of TA-ChNFs as both dynamic motifs and nanofillers significantly improved the mechanical properties of the St-based hydrogels. The highest stretchability of 1015% is observed for the 0.5 wt.% TA-ChNF-reinforced hydrogel. The highest tensile strength is seen for the 2 wt.% TA-ChNF-reinforced hydrogel (274.8 kPa). Moreover, the 1 wt.% TA-ChNF-reinforced hydrogel shows the highest toughness (1.43 MJ/m^3), which is 10.5-fold higher than the unreinforced hydrogel, whereas the decrease in toughness at higher TA-ChNFs may be due to a higher degree of crosslinking at higher TA-ChNFs concentrations.

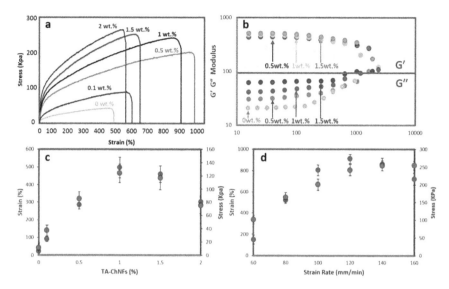

Figure 2. (**a**) Tensile stress–strain curves of PAA/St hydrogels at different TA-ChNFs concentrations; (**b**) Storage modulus and loss modulus of the hydrogel at different TA-ChNFs percentages; (**c**) tensile properties for notched hydrogels at different TA-ChNFs concentrations; (**d**) tensile properties of the hydrogel reinforced by 1 wt.% TA-ChNFs at different strain rates.

To evaluate the effects of different percentages of TA-ChNFs on the mechanical properties of the hydrogels, rheological measurements were performed. As seen in Figure 2b, the storage modulus (G′) of the hydrogel is higher than its loss modulus (G″), and the inclusion of TA-ChNFs, even at a very low concentration (0.5 wt.%), increases both the storage modulus (G′) and the loss modulus (G″) of the hydrogel; however, the trend is approximately the same at different TA-ChNFs concentrations, mainly due to both the nano-reinforcing and dynamic cross-linking effects of TA-ChNFs via non-covalent interactions. To assess the notch sensitivity of the TA-ChNF-reinforced hydrogels at different concentrations, all hydrogels were notched and stretched under tensile loading. It was observed that the hydrogel reinforced by 1 wt.% TA-ChNF had the highest notch-insensitivity, remaining remarkably stable and blunted. The toughness behavior of this hydrogel can be considered as the main reason for the resistance of hydrogel against crack propagation (Figure 2c). Based on the results from the mechanical, rheological, and notch-insensitivity measurements, the 1 wt.% TA-ChNF-reinforced hydrogel was considered as the optimum sample to pursue the remaining experiments.

The mechanical properties of the 1 wt.% TA-ChNF-reinforced hydrogel at different stretching rates (60–160 mm/min) was then tested to track the strain-rate dependency of the hydrogel. It was observed that the higher strain rates resulted in a higher breaking stress, reaching a maximum at 140 mm/min, beyond which the strain rate dependency of the tensile strength diminished. This can be attributed to a reduced dissipation energy by the coordination bonds [18]. While increasing the stretching rate up to 120 mm/min increased the breaking strain of the hydrogel, it decreased after 120 mm/min, mainly because of the inability of the broken physical bonds to reform (Figure 2d).

3.3. Morphological and FTIR Studies

In the next step, the morphological and FTIR studies of the optimized hydrogel were performed. Figure 3 shows the fracture surface of the pristine and 1 wt.% TA-ChNF-reinforced hydrogels. As can be seen, there exists a significant difference in the pore size and morphology of the reinforced and unreinforced hydrogels. While the reinforced hydrogel with 1 wt.% TA-ChNFs (Figure 3b) has a smaller, denser, and more uniform pore size, with a larger surface area, its unreinforced counterpart (Figure 3a) lacks any homogeneity and uniformity in pore size distribution. Therefore, it can be stated that the presence of TA-ChNFs enhances the homogeneity of the network by providing reversible interactions with the matrix. It can also be noted that the uniform structure of the 1 wt.% TA-ChNF-reinforced hydrogel resulted in better mechanical properties compared to its unreinforced counterpart.

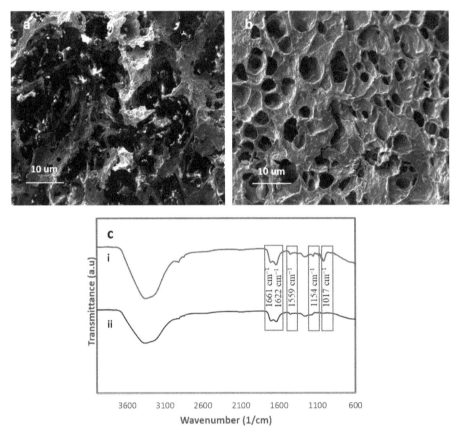

Figure 3. SEM images of (**a**) the unreinforced hydrogel and (**b**) the reinforced hydrogels with 1 wt.% TA-ChNFs; FTIR results of (**ci**) 1 wt.% TA-ChNF-reinforced hydrogels and (**cii**) unreinforced hydrogels.

Based on the FTIR results for the 1 wt.% TA-ChNF-reinforced hydrogel, there exist two amide I bands at 1622 and 1661 cm^{-1} and the amide II band at 1559 cm^{-1} (Figure 3ci,cii) due to the graft polymerization of AA on starch, and two characteristic bands at 1154 and 1017 cm^{-1} in Figure 3ci after the inclusion of 1 wt.% TA-ChNFs into the system. According to Ifuku et al. [29], these peaks are the C–O and C=O stretching vibration modes of the carboxylic acid resulting from the grafting of AA onto the chitin nanofibers.

3.4. Self-Healing and Self-Recovery (SELF) Properties

Using the rheological measurement, the self-recovery of the hydrogel containing 1 wt.% TA-ChNFs was demonstrated at the microscopic level. For this purpose, an alternate step strain test (strain = 1, 80, 300, 800, and 1000%) at a fixed time interval (100 s) was performed (Figure 4a,b). As can be seen, under small oscillatory strains, the hydrogel network is intact thanks to the presence of TA-ChNFs as dynamic motifs. It can be said that the polymer chains in the hydrogel are attached by TA-ChNFs; thus, they can bear a large deformation and show a rapid and complete self-recovery at a low strain of 1% due to the interactions between the nanofillers and matrix. This proves the excellent self-recovery of the hydrogel under repeatable cycles and the fact that the hydrogel could withstand large deformations without showing shear-thinning behavior, which stands in contrast with many self-healing hydrogels due to the presence of strong dynamic motifs created by TA-ChNFs [9,30,31]. At a 1000% strain, the value of G' and G" overlapped, which means the structure of the polymer hydrogel starts collapsing at such large strains (Figure 4b). When stepping the strain back to 1% at a fixed frequency (1.0 Hz), an instantaneous self-recovery to the gel-like character (G' > G") was observed in each repeatable cycle of the recovery. This observation can be attributed to a fully dynamic network that enables a response to the strain at a molecular level. The hydrogel also showed a shear-thinning behavior at strains higher than 1000%, which indicates the transformation of the hydrogel into its sol-gel state due to the destruction of the hydrogel network (Figure 4c).

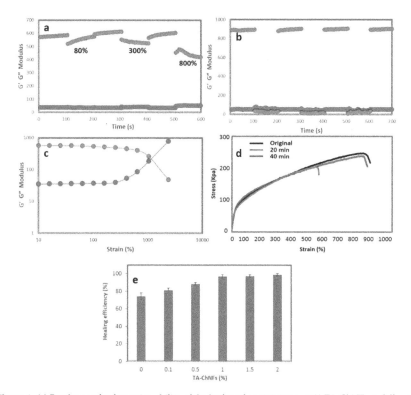

Figure 4. (**a**) Breakup and reformation ability of the hydrogel containing 1 wt.% TA-ChNFs at different strains = 1, 80, 300, and 800% and a fixed time interval; (**b**) breakup and reformation ability of the hydrogel containing 1 wt.% TA-ChNFs at 1 and 1000% strains and a fixed time interval; (**c**) shear-thinning behavior of the hydrogel containing 1 wt.% TA-ChNFs; (**d**) typical stress–strain curves of the pristine and healed hydrogels containing 1 wt.% TA-ChNFs; and (**e**) healing efficiency of the hydrogels at different TA-ChNFs.

To evaluate the self-healing performance of the hydrogel, a macroscopic test was performed using a direct visual inspection, followed by a tensile test. As shown in Video S3, a cylindrical-shaped hydrogel containing 1 wt.%. TA-ChNFs was cut into several pieces, and the pieces were then immediately rejoined together without applying any pressure. As seen, the hydrogel instantly healed itself without any external stimuli or healing agents. Moreover, the self-healed hydrogel showed immediate and stable self-support and stretchability across the cutlines due to the strong dynamic crosslinks. Clearly, the hydrogel showed a good healing ability, stability, and mechanical properties due to the presence of the rigid but dynamic TA-ChNFs motifs together with the coordination bonds.

To further investigate the effect of time on the healing efficiency of hydrogels, further tensile tests were performed by defining the healing efficiency as a tensile strength ratio between the healed and original hydrogel at the breaking point. As shown in Figure 4d, the hydrogel shows a time-dependent self-healing performance, and the tensile strength increases significantly by increasing the time, especially during the earlier step of healing. The self-healing efficiency reached 80% after 20 min, but after that it slowed down slightly and reached 96.5% after 40 min. Therefore, it can be said that this rapid and autonomous self-healing performance of the hydrogel is due to the presence of coordination crosslinks and TA-ChNFs dynamic motifs. Figure 4e shows the influence of different TA-ChNFs concentrations on the healing efficiency of the hydrogel. As can be seen, by adding 2 wt.% TA-ChNFs, the healing efficiency reached 98.5% after 40 min of healing. Therefore, the presence of TA-ChNFs plays a crucial role in the high SELF performance, network stability, and superior mechanical properties of the hydrogel due to providing dynamic metal-phenol networks.

3.5. Self-Adhesiveness Properties

The incorporation of TA into ChNFs, in addition to providing high SELF and mechanical properties, can impart a mussel-inspired adhesive mechanism to the hydrogels due to the presence of pyrogallol/catechol groups in TA, allowing the hydrogel to adhere to almost any surface [18,23]. As demonstrated in Video S4, the hydrogel containing 1 wt.% TA-ChNFs can merge the glass–plastic slides while lifting a load of 5 kg without using extra glue. It has been reported that the mussel-inspired adhesive mechanism of hydrogels depends on the availability of pyrogallol/catechol moieties and the type of interactions at the hydrogel–substrate interface, while the mechanical strength of the hydrogel depends on the number of TA/Fe^{3+} coordinations [32,33]. The self-adhesive properties of the hydrogel reinforced by 1 wt.% TA-ChNFs were quantified to different surfaces consisting of rubber, metal, glass, plastic, and leather using a tensile test. As Figure 5a depicts, the highest self-adhesive strength was found between hydrogel–metal interfaces. This can be attributed to the presence of metal complexation and hydrogen bonding at the interfacial layer [23]. The adhesiveness of hydrogel–plastic can be attributed to hydrophobic interactions, e.g., π-π stacking or CH-π interaction. Hydrogen bonding can also be considered for the adhesiveness of the hydrogel to rubber, glass, and leather. [18] A cyclic peel-off test was also performed four times by adhering 1 wt.% TA-ChNF-reinforced hydrogel onto different substrates and peeling them off by the tensile load, followed by re-adhering them onto the same substrates and repeating the test (Figure 5b). Based on the results, the hydrogel showed good repeatable self-adhesive properties. As an example, 76% of the initial self-adhesive strength between the hydrogel–metal at the first cycle was restorable after the fourth cycle. As Video S5 shows, the hydrogel can easily adhere to rubbery gloves while being stretched without adding any additional adhesive tapes. Therefore, the added functionality of self-adhesiveness to the hydrogel can, without influencing the mechanical performance and self-healing, increase the versatility of the hydrogel in many practical applications, e.g., wearable strain sensors [32].

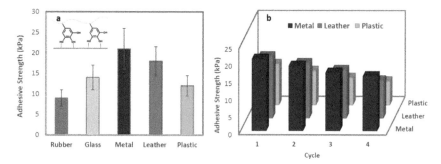

Figure 5. (**a**) Self-adhesive strength of the 1 wt.% TA-ChNF-reinforced hydrogel on different substrates measured by a tensile test; (**b**) cyclic self-adhesive tests on different substrates.

3.6. Electrical Conductivity and a Potential Application

The electrical conductivity is demonstrated by a light-emitting diode (LED) indicator and the hydrogel reinforced by 1 wt.% TA-ChNFs as the conductor in Videos S6 and S7. As shown in Video S6, the LED bulb reversibly darkened and lit up by applying and releasing the stress due to the presence of local disconnections, indicating the fast resistance response of the hydrogel to the strain. The soft electrical switching behavior of the hydrogel was observed using the LED indicator and is shown in Video S7. As can be seen, the LED bulb in the electric circuit was lit by connecting the hydrogel to the electric circuit. The hydrogel was then cut in half using scissors, and the cut pieces were again connected. As seen in Figure 6a,b, The LED bulb in the electric circuit was instantly lit, and the circuit was restored due to the migration of free Fe^{3+} ions from one side to another (Figure 6c), imparting a restorable ionic conductivity to the hydrogel. The resistance of the hydrogel was measured using a multimeter at a probe distance of 1 cm. It was observed that by increasing the strain values from 0 to 600%, the resistance of the hydrogel increased. This can be attributed to the increased distance between the conductive segments of the hydrogel network, resulting in increased local disconnections within the network (Figure 6d) [23].

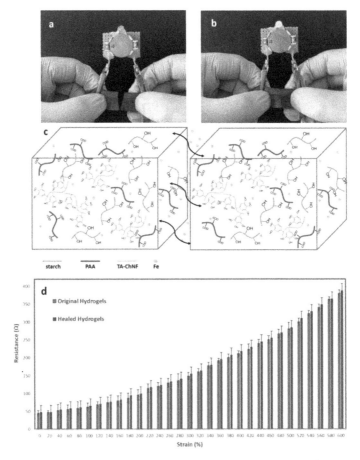

Figure 6. (**a**) Before connecting the cut hydrogel containing 1 wt.% TA-ChNFs; (**b**) after connecting the cut hydrogel with 1 wt.% TA-ChNFs; (**c**) the mechanism of electrical self-healing of the hydrogel and (**d**) the resistance of the hydrogel at different strains.

Most hydrogel-based wearable sensors require an external glue for affixing to the body. Moreover, they are usually brittle and are likely to lose their functionality while under load [18]. By taking advantage of TA-ChNFs, a self-adhesive hydrogel with the potential to solve this problem can be fabricated. The potential use of the hydrogel as a self-wearable strain sensor was examined by adhering it onto the top surface of the forefinger without employing any glue for detecting the bending movement of the finger. As Figure 7 depicts, the hydrogel could be easily attached to the forefinger without using any glue, and by increasing the angle of bending from 0° to approximately 90° the resistance of the hydrogel increased, whereas the resistance did not change when the forefinger was held static at approximately the same angle. Interestingly, by returning the bending angle to 0°, the initial resistance of the hydrogel was restored. Hence, the hydrogel can distinguish between different bending angles based on the relative resistance changes in real time, making our hydrogel an ideal candidate for self-wearable strain sensors.

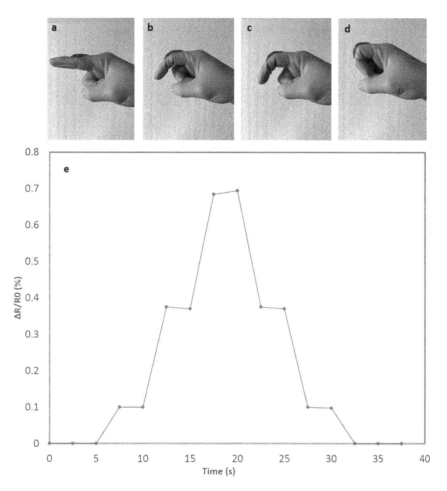

Figure 7. (**a–d**) Adhered 1 wt.% TA-ChNFs hydrogel to the first author's forefinger while bending to approximate angles and (**e**) its relative resistance change in real time.

4. Conclusions

In this work, a dynamic mussel-inspired hydrogel was designed and fabricated by incorporating Fe^{3+} ions and TA-ChNFs into a starch-based network. TA-ChNFs play the role of nanofillers and dynamic cross-linkers, thus imparting a SELF ability and high strength to the hydrogel. The hydrogel shows a high SELF ability, mechanical strength, and electro-conductivity. According to our results, the hydrogel reached the highest toughness and notch insensitivity after loading 1 wt.% TA-ChNFs into the hydrogel. Moreover, the hydrogel reinforced by 2 wt.% TA-ChNFs showed the highest self-healing efficiency (98.5%) after 40 min. Additionally, the hydrogel showed repeatable self-adhesive properties, with the ability to attach to almost any surface because of the existence of pyrogallol/catechol groups in TA. The hydrogel is also able to monitor and distinguish motions, showing a good capability as a soft wearable strain sensor. Thus far, there has not been any available research on using TA-ChNFs and starch as potential candidates for fabricating such hydrogels. We anticipate that the hydrogel will be ideally suited for use in self-wearable flexible strain sensors because of its excellent self-healing ability—without needing any external stimuli—and reliable mechanical, electrical, and self-adhesive properties.

Supplementary Materials: The following are available online at http://www.mdpi.com/2073-4360/12/6/1416/s1,. Video S1: 0.5 wt.% TA-ChNF-reinforced hydrogel. Video S2: 0.1 wt.% TA-ChNF-reinforced hydrogel. Video S3: Visual self-healing inspection of a cylindrical-shaped 1 wt.%. TA-ChNF-reinforced hydrogel. Video S4: Merging glass-plastic slides by a 1 wt.%.TA-ChNF-reinforced hydrogel. Video S5: 1 wt.%.TA-ChNF-reinforced hydrogel adhered to rubbery gloves while being stretched without adding any additional adhesive tapes. Video S6: Darkening and lightening of 1 wt.%.TA-ChNF-reinforced hydrogel after applying and releasing the stress using an LED bulb indicator. Video S7: Electrical self-healing ability of the hydrogel reinforced by 1 wt.%.TA-ChNF.

Author Contributions: Conceptualization, methodology, acquisition of data, analysis and interpretation of data, drafting of the manuscript, P.H.; project administration, supervision, resources, review and editing, A.Z.K.; supervision, discussions, analysis and interpretation of data, review and editing, A.K.; review and editing, A.Z.; review and resources, H.Y. All authors have read and agreed to the published version of the manuscript.

Funding: This research received no external funding.

Conflicts of Interest: The authors declare no conflict of interest.

References

1. Heidarian, P.; Kouzani, A.Z.; Kaynak, A.; Paulino, M.; Nasri-Nasrabadi, B.; Zolfagharian, A.; Varley, R. Dynamic Plant-Derived Polysaccharide-Based Hydrogels. *Carbohydr. Polym.* **2020**, *231*, 115743. [CrossRef] [PubMed]

2. Heidarian, P.; Kouzani, A.; Kaynak, A.; Paulino, M.; Nasri-Nasrabadi, B. Dynamic Hydrogels and Polymers as Inks for 3D Printing. *ACS Biomater. Sci. Eng.* **2019**, *5*, 2688–2707. [CrossRef]

3. Deng, Z.; Wang, H.; Ma, P.X.; Guo, B.J.N. Self-healing conductive hydrogels: Preparation, properties and applications. *Nanoscale* **2020**, *12*, 1224–1246. [CrossRef] [PubMed]

4. Zhang, Y.S.; Khademhosseini, A.J.S. Advances in engineering hydrogels. *Science* **2017**, *356*, eaaf3627. [CrossRef]

5. Larson, C.; Peele, B.; Li, S.; Robinson, S.; Totaro, M.; Beccai, L.; Mazzolai, B.; Shepherd, R. Highly stretchable electroluminescent skin for optical signaling and tactile sensing. *Science* **2016**, *351*, 1071–1074. [CrossRef]

6. Nasri-Nasrabadi, B.; Kaynak, A.; Heidarian, P.; Komeily-Nia, Z.; Mehrasa, M.; Salehi, H.; Kouzani, A. Sodium alginate/magnesium oxide nanocomposite scaffolds for bone tissue engineering. *Polym. Adv. Technol.* **2018**, *29*, 2553–2559. [CrossRef]

7. Basu, A.; Lindh, J.; Ålander, E.; Strømme, M.; Ferraz, N. On the use of ion-crosslinked nanocellulose hydrogels for wound healing solutions: Physicochemical properties and application-oriented biocompatibility studies. *Carbohydr. Polym.* **2017**, *174*, 299–308. [CrossRef]

8. Zolfagharian, A.; Kouzani, A.Z.; Khoo, S.Y.; Nasri-Nasrabadi, B.; Kaynak, A. Development and analysis of a 3D printed hydrogel soft actuator. *Sens. Actuators Phys.* **2017**, *265*, 94–101. [CrossRef]

9. Taylor, D.L.; in het Panhuis, M.J.A.M. Self-healing hydrogels. *Adv. Mater.* **2016**, *28*, 9060–9093. [CrossRef]

10. Kowalski, P.S.; Bhattacharya, C.; Afewerki, S.; Langer, R. Smart biomaterials: Recent advances and future directions. *ACS Biomater. Sci. Eng.* **2018**, *4*, 3809–3817. [CrossRef]

11. Ge, W.; Cao, S.; Shen, F.; Wang, Y.; Ren, J.; Wang, X. Rapid self-healing, stretchable, moldable, antioxidant and antibacterial tannic acid-cellulose nanofibril composite hydrogels. *Carbohydr. Polym.* **2019**, *224*, 115147. [CrossRef] [PubMed]

12. Nadgorny, M.; Ameli, A. Functional polymers and nanocomposites for 3D printing of smart structures and devices. *ACS Appl. Mater. Interfaces* **2018**, *10*, 17489–17507. [CrossRef] [PubMed]

13. Qin, Y.; Wang, J.; Qiu, C.; Xu, X.; Jin, Z. Dual Cross-Linked Strategy to Construct Moldable Hydrogels with High Stretchability, Good Self-Recovery, and Self-Healing Capability. *J. Agric. Food Chem.* **2019**, *67*, 3966–3980. [CrossRef]

14. Wu, X.; Wang, J.; Huang, J.; Yang, S. Robust, stretchable, and self-healable supramolecular elastomers synergistically cross-linked by hydrogen bonds and coordination bonds. *ACS Appl. Mater. Interfaces* **2019**, *11*, 7387–7396. [CrossRef] [PubMed]

15. Shao, C.; Meng, L.; Wang, M.; Cui, C.; Wang, B.; Han, C.-R.; Xu, F.; Yang, J. Mimicking dynamic adhesiveness and strain-stiffening behavior of biological tissues in tough and self-healable cellulose nanocomposite hydrogels. *ACS Appl. Mater. Interfaces* **2019**, *11*, 5885–5895. [CrossRef] [PubMed]

16. Liu, Q.; Zhang, M.; Huang, L.; Li, Y.; Chen, J.; Li, C.; Shi, G. High-quality graphene ribbons prepared from graphene oxide hydrogels and their application for strain sensors. *ACS Nano* **2015**, *9*, 12320–12326. [CrossRef] [PubMed]

17. Zhong, M.; Liu, Y.-T.; Xie, X. Self-healable, super tough graphene oxide–poly (acrylic acid) nanocomposite hydrogels facilitated by dual cross-linking effects through dynamic ionic interactions. *J. Matter. Chem. B* **2015**, *3*, 4001–4008. [CrossRef]

18. Shao, C.; Wang, M.; Meng, L.; Chang, H.; Wang, B.; Xu, F.; Yang, J.; Wan, P. Mussel-inspired cellulose nanocomposite tough hydrogels with synergistic self-healing, adhesive, and strain-sensitive properties. *Chem. Mater.* **2018**, *30*, 3110–3121. [CrossRef]

19. Jayakumar, R.; Nwe, N.; Tokura, S.; Tamura, H. Sulfated chitin and chitosan as novel biomaterials. *Int. J. Biol. Macromol.* **2007**, *40*, 175–181. [CrossRef]

20. Dutta, P.K.; Dutta, J.; Tripathi, V. Chitin and chitosan: Chemistry, properties and applications. *J. Sci. Ind. Res.* **2004**, *63*, 20–31.

21. Hejazi, M.; Behzad, T.; Heidarian, P.; Nasri-Nasrabadi, B. A study of the effects of acid, plasticizer, cross-linker, and extracted chitin nanofibers on the properties of chitosan biofilm. *Compos. Part A-Appl.* **2018**, *109*, 221–231. [CrossRef]

22. Mohammadinejad, R.; Maleki, H.; Larrañeta, E.; Fajardo, A.R.; Nik, A.B.; Shavandi, A.; Sheikhi, A.; Ghorbanpour, M.; Farokhi, M.; Govindh, P. Status and future scope of plant-based green hydrogels in biomedical engineering. *Appl. Mater. Today* **2019**, *16*, 213–246. [CrossRef]

23. Heidarian, P.; Kouzani, A.Z.; Kaynak, A.; Paulino, M.; Nasri-Nasrabadi, B.; Varley, R. Double dynamic cellulose nanocomposite hydrogels with environmentally adaptive self-healing and pH-tuning properties. *Cellulose* **2019**, *27*, 1407–1422. [CrossRef]

24. Shen, X.; Nie, K.; Zheng, L.; Wang, Z.; Wang, Z.; Li, S.; Jin, C.; Sun, Q. Muscle-inspired capacitive tactile sensors with superior sensitivity in an ultra-wide stress range. *J. Mater. Chem. C.* **2020**, *8*, 5913–5922. [CrossRef]

25. Chen, H.; Gao, Y.; Ren, X.; Gao, G. Alginate fiber toughened gels similar to skin intelligence as ionic sensors. *Carbohydr. Polym.* **2020**, *235*, 116018. [CrossRef]

26. Hussain, I.; Sayed, S.M.; Liu, S.; Oderinde, O.; Yao, F.; Fu, G. Glycogen-based self-healing hydrogels with ultra-stretchable, flexible, and enhanced mechanical properties via sacrificial bond interactions. *Int. J. Biol. Macromol.* **2018**, *117*, 648–658. [CrossRef] [PubMed]

27. Fan, H.; Wang, J.; Jin, Z. Tough, Swelling-Resistant, Self-Healing, and Adhesive Dual-Cross-Linked Hydrogels Based on Polymer–Tannic Acid Multiple Hydrogen Bonds. *Manromolecules* **2018**, *51*, 1696–1705. [CrossRef]

28. Heidarian, P.; Behzad, T.; Sadeghi, M. Investigation of cross-linked PVA/starch biocomposites reinforced by cellulose nanofibrils isolated from aspen wood sawdust. *Cellulose* **2017**, *24*, 3323–3339. [CrossRef]

29. Ifuku, S.; Iwasaki, M.; Morimoto, M.; Saimoto, H. Graft polymerization of acrylic acid onto chitin nanofiber to improve dispersibility in basic water. *Carbohydr. Polym.* **2012**, *90*, 623–627. [CrossRef]

30. Li, Q.; Liu, C.; Wen, J.; Wu, Y.; Shan, Y.; Liao, J. The design, mechanism and biomedical application of self-healing hydrogels. *Chin. Chem. Lett.* **2017**, *28*, 1857–1874. [CrossRef]

31. Zhao, H.; An, H.; Xi, B.; Yang, Y.; Qin, J.; Wang, Y.; He, Y.; Wang, X. Self-healing hydrogels with both LCST and UCST through cross-linking induced thermo-response. *Polymers* **2019**, *11*, 490. [CrossRef] [PubMed]

32. Fan, H.; Wang, J.; Zhang, Q.; Jin, Z. Tannic acid-based multifunctional hydrogels with facile adjustable adhesion and cohesion contributed by polyphenol supramolecular chemistry. *ACS Omega* **2017**, *2*, 6668–6676. [CrossRef] [PubMed]

33. Sahiner, N.; Butun Sengel, S.; Yildiz, M.J.J.o.C.C. A facile preparation of donut-like supramolecular tannic acid-Fe (III) composite as biomaterials with magnetic, conductive, and antioxidant properties. *J. Coord. Chem.* **2017**, *70*, 3619–3632. [CrossRef]

MDPI

St. Alban-Anlage 66

4052 Basel

Switzerland

Tel. +41 61 683 77 34

Fax +41 61 302 89 18

www.mdpi.com

Polymers Editorial Office

E-mail: polymers@mdpi.com

www.mdpi.com/journal/polymers